Animals & Men
#71

Contents

Typeset by some small bantams
Cover and Layout by SPiderKaT for CFZ Communications
Using Microsoft Word 2000, Microsoft Publisher 2000, Adobe Photoshop CS.
First published in Great Britain by CFZ Press

CFZ Press, Myrtle Cottage, Woolsery, Bideford, North Devon, EX39 5QR

© CFZ MMXXIII

ISBN: 9798377910374

Faculty of the Centre for Fortean Zoology

Dear Friends,

Welcome to a very belated issue of the English speaking world's longest standing magazine dealing with Cryptozoology and allied disciplines. I remember thirty plus years ago when I was editing a music magazine that nobody bought (no change there then), that I was sent a fanzine put together by a well meaning and earnest schoolboy. His editorial opened with apologies that it has been so long since the previous issue, but he went on to say the reason that the issue had been so badly delayed was that he had been cast to appear in his school production of "Oklahoma!" which I found massively funny, for unknown reason.

I have a better, but also far more sad excuse: My wife Corinna, whom many of you knew from the Weird Weekend and from her 'Watcher of the Skies' column in this magazine, died of cancer in the summer of 2020, and I have been dealing with the fallout, emotional, physical and all sorts of other things.

It was not unexpected, because she had been very ill for a long time. However, for reasons which I don't really understand, she didn't want anybody to know this and so it came as a great shock to many people when I announced that she had transitioned to the next world and joined her ancestors. Corinna was not only my wife, but for the last fifteen years she had also been the administrative director of the Centre for Fortean Zoology. So, as I am sure you can imagine, I have not only been mourning my wife, but sorting out the administrative hierarchy of the CFZ.

To make things worse, I have also managed to severely damage my feet, and as a result I have been in and out of hospital anything up to four times a week.

On a positive note, we are well on our way now to resuming our normal service (or whatever is "normal" under the strange circumstances in which we all find ourselves these days).

We are continuing at pace with our publication schedule, and I am happy to announce that we have published the first English language edition of Boris Porshnev's legendary 1963 book on relict hominins in the Soviet Union. Marie-Jeanne Koffmann once explained her theory as to why the Soviet Union, even at its most repressive under Stalin, was the only country in the world to

The Great Days of Zoology are not done!

have a state sponsored organisation whose mission was to look for unknown animals. She believed that official party line was that as communism functioned through a doctrine of collectivism, the fact that there were bipedal hominids living in the Soviet Union which had not passed through what I have always called the 'Civilization Threshold' proved that bipedalism and opposable thumbs were not responsible for human civilization. I personally think this is a load of nonsense, but who am I to second guess Stalin?

You will find more details of the Porshnev book at the end of this issue.

You also may have already heard the exciting news that Richard and I have negotiated the purchase of a cache of papers from the estate of the late Odette Tchernine.

You may know that Odette Tchernine was a polymath, who wrote on a number

of subjects including science fiction, but her publications include three books on relict hominins. The papers that we have purchased include her unpublished fourth book, and I am proud to be able to tell you that we intend to publish it sometime next year, as the work required to put this book together has already commenced!

However, whilst perusing the rest of her papers, we became impressed by her polymathic output and have decided to attempt to celebrate the life of this woman, who most likely otherwise would fade from our collective memories. Guin Palmer is collecting more material from newspapers and various other archives, which we intend to use in an application for an Arts Council grant which will enable us to fund the development of a WebApp/Website honouring her memory. All sorts of things are coming out of the woodwork, and we intend to feature tributary videos from people who knew her including Colonel John Blashford-Snell and Tony Healy.

I hope that you agree with me that this is a worthwhile project with which we should certainly be involved. I am also pleased to announce that the legendary book by Boris Porshnev, about Soviet era wild men, is now fully available for sale and there are details elsewhere in this newsletter.

Carl Marshall and his merry band are continuing their work in the Forest of Dean. It is too early to be able to give you any definitive conclusions, but from where I am sitting, the existence of a population of medium sized exotic cats such as Lynx or jungle cats seems extremely likely. In

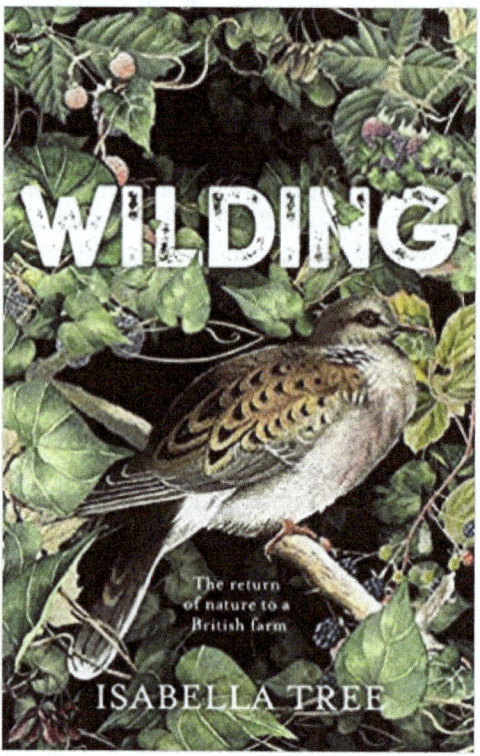

The premise of the book is two-fold: That rewilding an area like Knapp Castle, where Tree and her husband Sir Charles Burrell live, and have been carrying out an intensive program of the same, does not necessarily have to involve introducing fearsome wild animals. The Knapp Castle project has proven that one of the best things that one can do is just leave an area be. Cuckoo's, Turtle Doves, Purple Emperors and a plethora of other species have recolonised lands in which they once lived surprisingly quickly once the land has been left to fall fallow.

The second part of their thesis concerns their discovery, that contrary to much accepted thought, the primordial forest cover of Great Britain was almost certainly not thick canopy growth; the introduction of Ponies, Pigs and Cattle of as closer breeds as could be managed to the wild horses, wild boar and aurochs which once lived across North-western Europe, has fostered a well-drained but surprisingly varied terrain with a wide range of habitats for different creatures.

If nothing else, this is a biogeographical discovery which turns much of what we thought to be true about pre-modern British topology on its head.

It is books like this that encourage us all to question the established paradigms, which is - of course - something which scientists and Forteans should do constantly.

The October before last one of my favourite contemporary natural history writers, Patrick Barkham wrote a fascinating article in The Guardian in which he described the activities of the self styled "Introductionists". These people are

his books, Dr Karl Shuker has presented conclusive evidence (he has one of the bodies in his possession) that there have been jungle cats living in the wilds of Shropshire within the past forty years and, something that makes this even more intriguing, he has presented a reasonable amount of evidence to suggest that the jungle cat had interbred with some of the local population of feral moggies.

The other summer Max Blake bought me a copy of 'Wilding' by Isabella Tree for a birthday present. I was immediately entranced and have bought copies for various people I know, most notably Carl Marshall's Dad who was completely bowled over by it.

responsible, or so it seems, for the reintroduction of many species of animal which have been presumed lost, into the British countryside in recent years. The article concentrated on the work of Martin White who had died of cancer by the time the article was published. White's final project was to reintroduce Mazarine Blues, a species which became extirpated in these islands in the years leading up to the first world war. The article, should you like to read it, can be found here:

https://www.theguardian.com/ environment/2020/oct/13/maverick-rewilders-endangered-species-extinction-conservation-uk-wildlife

It also describes how all sorts of animals from Polecats to Beavers have been reintroduced by these well meaning, but completely illegal, amateurs. My opinion on these people is that largely they are a force for good. I think that it is incontrovertible that British government legislation involving anything to do with the environment, is needlessly complicated and often badly thought out and counterproductive. The Dangerous Wild Animals Act of 1976 was a ludicrous thought out piece of legislation, which resulted in all sorts of alien species being introduced to the British countryside, purely because the new legislation which had to be obeyed by people already keeping these creatures as domestic pets or as a part of research projects, was so draconian that the owners were faced with very little option but to release or destroy them. I know which option I would have chosen.

However, it is also incontrovertible that with inadequate biosecurity, reintroduction projects can end up causing carnage like the Chytrid fungus - *Batrachochytrium dendrobatidis*, which it has been suggested has spread across the world with the trade in African clawed frogs (*Xenopus laevis*) and American Bullfrogs (*Lithobates catesbeianus*).

So, cautiously, I welcome the activities of these people, but I am totally aware that this statement may come back to bite me in the bum.

We are also very proud to say that this issue is the first one to have been issued in hardcopy for many years. Both Richard, Freeman and I, I have been very upset that we have not been in a position to be able to issue the magazines in any other format than electronically. I am perfectly aware that the vast majority of people prefer to consume the magazines and the journals in an electronic format, therefore, we will still be issuing the magazines electronically through Flipsnack, embedded on the website, but for those of us who are old-fashioned enough to want magazines in hardcopy, you will be able to find them on Amazon.

I hope you enjoy this issue, and you continue to support our various endeavours.

Yours, as ever,

Jonathan Downes (Director, CFZ)

THE CENTRE FOR FORTEAN ZOOLOGY
www.cfz.org.uk

A LEGAL MATTER

Community Notices

The newsletter is now free

Like many people my age I watched the movie of Woodstock, partly because of all the naked girls and partly because it had Jimi Hendrix in it. However, about three quarters of the way through the movie, around about the time The Who did their thing, Michael Lang climbed onto the stage and told everybody that the festival was now going to be free. I always felt that if I had been someone who had already paid for a ticket, I would have been mightily annoyed.

Michael Lang's hand had been forced by the fact that tens of thousands of people had ripped down the perimeter fence and got in for nothing. I have no such excuse.

Louis, who most of you know as producer of On The Track and our digital presence coordinator, has told me to stop charging for this newsletter. So, from now on this is going to be a free festival, umm newsletter. Please don't be too angry with me, because I have to go on stage now and play the Star Spangled Banner with a load of feedback.

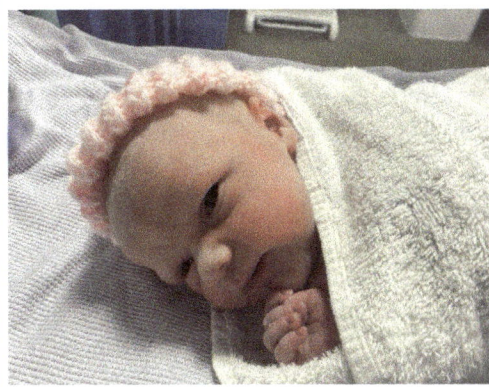

Welcome Elsie

Regular readers of my inky fingered scribblings, here and elsewhere, will know David and Jessica Braund-Phillips. I have known David all his life, and it has been my privilege to have been Uncle Jon since he was born. I have known Jessica for 12 or 13 years, and it is my great pleasure to announce a CFZ sprog: my great-niece Elsie, whom I am sure is going to be running about in various costumes for On the Track as soon as she is able to run about. Congratulations my darlings.

Changes to the website and OTT.

Regular CFZ watchers will have, no doubt, noticed that the Centre for Fortean Zoology website has been given its biggest overhaul in the last 20 years.

Until now it has been managed by Graham and Jon, but with the advances in the way that the Internet works, they were no longer able to produce something as sophisticated as the market required. Therefore, despite their best efforts, it was getting more and more out of date.

Earlier this year, Jon's friend Louis Rozier Who – a few years ago – set up the Patreon page for our web TV show, offered to make a new website for us, and he was as good as his word. The new website is completely magnificent, and considerably easier to use and update than its predecessors. We cannot thank him enough.

He has also taken over as producer of our web TV show On the Track which is now weekly. I am sure you will agree that the production values which were a little shoddy at times have improved hand over fist, and that we are going from strength to strength.

Charlotte Phillipson has left the core team after three or four years because she is currently at Aberystwyth University studying marine biology.

We wish her all the luck in the world, and she is the recipient of all our love and positive vibes and any time she turns up here to see us we will probably, nay almost certainly, video her doing a brief cameo or two.

And finally…. more babies.

The eagled eyed amongst us will know that in the past few years Guin Palmer (often referred to as Miss Guinevere) has become ever more important whilst the CFZ adjusts to its latest iterations. Indeed, she has taken over from my late wife as the CFZ administratrix.

As well as carrying out these and other important actions, Miss Guinevere has also recently brought two little boys, to whom I am Great Uncle Jon, into the world.

So, Ladies and Gentlemen, herewith I proudly present my nephews Muhammed (L) and Ibrahim (R).

CORINNA DOWNES
(1956-2020)

In March 2005 I met, and started going out with, a lady called Corinna James.

Unusually for most of the women I have known, she had impeccable taste in music, particularly progressive rock. In the speech I made at our wedding two years later I only half joked that I had to marry her because she was the only woman I've ever met who had ever seen Henry Cow live. She had two daughters, Olivia and Shoshannah, born in the 1980s, and - gratifyingly - they were very kind to me and welcomed me into their family when I became their stepfather.

The CFZ welcomed her with open arms and she was soon much loved. She became the Administrative Director in 2006, and managed to make some sort of sense of the chaos caused by me, Graham and Richard over the years. In 2007 she moved to be with me in North Devon and in July that year we got married, and were very happy.

That Christmas her father died by his own hand, and that is the sort of thing you never recover from. A year or two later somebody whom I have introduced to her house trustworthy turned out not to be and took £50,000 of her savings. I'm not going to go into any more details, because the matter is now in the hands of solicitors. She never recovered fully from these two blows, and over the years she became more withdrawn, which is why events like the Weird Weekend, and other business/social functions were truncated.

But we were happy in ourselves, and I was glad to put some of my more high-profile cryptozoology activities on the back burner for her sake. In 2013 her mother, whom I loved very much, moved into the CFZ library and stayed with us for the rest of her life, dying in 2019.

In 2014 my beloved younger stepdaughter Olivia gave birth to a little girl called Evelyn. Corinna and I were overjoyed to be grandparents. In 2018 Corinna was diagnosed with terminal cancer, and fought it valiantly. She didn't want anybody outside her close circle to know about this, and I respected her wishes as far as I could. But, as her condition progressed, and she became more ill, it was harder and harder to do this, when people from the CFZ family around the world were concerned with her obviously feeble condition. She died at 6:30 in the morning on the 16th of August 2020. My two stepdaughters and I were holding her hand as she passed to join her ancestors. My life will never be the same again, and although I do not intend to be one of these widowers who make their whole life into a shrine to their dead partner, whatever happens to me in the years to come, she will always have an enormous piece of my heart.

WOLFGANG SCHMIDT (195?-2020)

I first met Wolfgang at the first Fortean Times Unconvention in 1994. We bonded in the bar over cryptozoology and lager.

Over the years we became firm friends, and he was a generous benefactor to the Centre for Fortean Zoology. His most generous donation took place in 1999 when he gave us £1000 to buy camera equipment.

He visited us in Exeter for the third Weird Weekend in 2002, and we always met up at Unconventions where he seemed to do his best to buy the entire contents of our stall.

About 20 years ago he told me about, what seemed to me to be, a ridiculous scheme to build a complex of conference facilities and museums in the shape of pyramids. But bloody hell, a few years later he did just that.

He and I were always conspiring to get the Centre for Fortean Zoology to do a major exhibition at Galileo Park, which is what his pyramids are called, but always, something seemed to get in the way, most recently that 'something' being my family obligations. But we always thought we would get around to it eventually. But then came COVID-19, the western world went into lockdown, and despite all the precautions, Wolfie died of it last winter.

My life truly will be a poorer place without him.

NADINE 'DEANIE' RIDER (2001-2020)

Regular CFZ watchers will remember that, for many years, the CFZ had a whole string of volunteers and interns who also happened to be – conversely – charming young women. They became known as the CFZ Girls. Many of you will remember Nadine Rider, better to known to everyone as

'Deanie'. I am very unhappy to have to tell you that she died by her own hand in September 2020. The circumstances surrounding her death are unclear, but sufficient to say that we are all devastated.

Our love and prayers go out to her fiancé and family.

Michael Newton
1951-2021

MICHAEL NEWTON (1951-2021)

We regret to announce the death of Michael Newton, founding editor of CFZ-USA and prolific writer on cryptozoology and other subjects. We understand Michael died peacefully, with his wife and cats at his side.

Michael wrote his own obituary (below):

MICHAEL NEWTON

September 16, 1951-September 6, 2021.

Michael was born in Bakersfield, California in September 1951. He cultivated dreams of a writing career from age seven, producing small "books" illustrated with photos and original drawings, given or sold for a pittance to classmates. On the practical side, he earned a B.A. in History and Political Science, with a teaching minor in English, pursuing sundry jobs in public education in California (1973-76) and later in Nevada (1979-86).

Nearly by accident, Michael advanced his writing dream in 1976, penning a fan letter to the author of his favourite action/adventure novels, the best-selling "Executioner" series. He had forgotten that whimsical letter two months later, when series author Don Pendleton phoned from his home in Indiana, inviting Michael to contribute a substantial chapter to his latest work, The Executioner's War Book. That entry became Michael's first professional publication and earned him an invitation—with colleague Stephen Mertz, author of many books under sundry pen names—to work with Pendleton in Indiana. There ensued a priceless eight-month apprenticeship, during which Michael also cultivated a love for the verdant hills of Brown County. After that partnership dissolved in August 1977, Michael published eleven books under his own and various pen names. In 1980 he received a startling letter from Harlequin Books, thanking him for his interest in their new Gold Eagle action/adventure imprint, poised to revive and vastly expand the briefly moribund Executioner series. Divining that Don Pendleton had placed his name "in the hat," Michael auditioned for the team and won a spot as one of four. By the time Don passed away in 1995, Michael had penned fifty series episodes. He went on to publish a total of 136.

Michael met his best friend and soul mate, Heather in 2000, and they married at Fort Augustus, Scotland, on the shore of Loch Ness, in 2003.

Diagnosed with an untreatable hereditary kidney disease in 1988, Michael lived normally until declining health forced him into home dialysis in 2013. From there, he gained a new appreciation of two favourite singers: Mick Jagger ("What a drag it is getting old") and Jim Morrison ("No one here gets out alive").

As of 2021 Mike had published 357 books (a tribute to the .357 Magnum pistols, perhaps?) which included 258 novels and 99 nonfiction books. He also published 91 nonfiction articles, and 58 shorter pieces, including chapters in several best-selling true-crime anthologies. In 2017 Michael received the Lifetime Achievement Peacemaker Award from Western Fictioneers, honouring his publication of 62 western novels.

If any form of consciousness remains, he said he'd miss Heather, their cats, writing and reading. In lieu of flowers, Mike would love a PayPal donation to go to his favourite cat charity, Mara's Heaven, an amazing non-profit in Romania, run by Ada Constantinescu. (PayPal: ada_angel_77@yahoo.com) Check out her FB page and you'll see the amazing work she does.

C'est la vie, and adios y'all.

SCOTT MARDIS (1964-2021)

I hear with sadness that Scott Mardis, veteran Lake Champlain researcher, died suddenly. Apparently, he got an unexpected infection in one of his legs, to which he paid no attention until it started causing him severe pain. They took him into hospital and amputated his leg. However, he died a few days later.

We never met, but exchanged cordial emails on occasion.

The cryptozoological universe is going to be a considerably poorer place without him. Our heartfelt sympathies go out to his family and friends.

MARIE-JEANNE KOFFMANN (1919-2021)

Some time ago I heard from a supposed researcher, that they were giving up on cryptozoology because in the three years they had been in the field they had made no money and found nothing! Marie Jeanne-Koffman spent four decades in the mountains and forests of the former USSR in search of the Russian wild man known as the almasty.

Born in Paris in Paris on 19th of July 1919 Marie Jeanne-Koffman spent most of her life in Russia. She obtained an M.D. at Moscow University in 1941 and became a surgeon in Moscow. In her spare time became interested in Mountaineering. Marie served in the Red Army during World War II,

attaining the rank of captain. She fought in the Battle of Moscow, as well as in the Battle of the Caucasus, during which she was second-in-command of a battalion of mountain rangers. She received seven Soviet battle citations.

After the war she participated in the first expeditions to previously unexplored mountain ranges, including the Pamirs in 1947.

Later she was accused of being a spy for France she was held prisoner in a gulag from 1948 to 1954. She was eventually acquitted and released.

In 1957 Marie saw an article on the yeti in a Russian magazine. It covered the early expeditions and had opinions from mountaineers, several of whom she knew. This piqued her interest and she began her own research that carried on well into the 1990s. In 19S8, she was elected to the Society of Geography of the USSR Academy of Sciences.

She joined the Soviet Union Snowman Commission, a group of scientists based at the Darwin Museum in Moscow and dedicated to researching these creatures. The Commission members included scientists like Professor Pitor Smolin, Professor A.A. Machkovtsev and Dmitri Bayanov.
Her first field work was in the Pamir Mountains as the doctor for the first Snowman Commission expedition. Later she concentrated on Caucasus.

Dr. Koffmann spent decades in the Caucasus mountains searching for the almasty, as the locals called the wild man. She glimpsed the creature once from a distance herself and interviewed hundreds of witnesses and gathered copious notes on the creature's habits. Her impressive and substantial body of

work remains mostly unpublished. She also established the Russian Society of Cryptozoologists based at the Darwin Museum in Moscow.

Despite never writing a book on the almasty she **published a synthesis of her fieldwork and research in the French journal Archeologia. This included the creature's appearance, diet, behaviour and tool use.**

She was active in the field into the 1990s. Gregory Panchenko, the Ukrainian biologist who was the guide to the CFZ's almasty expedition in 2008 worked with her on some of her later trips.

Dr. Koffmann left Russian in 2009 at the age of 90. She lived at the **Gautier Wendelen** retirement home. Despite suffering a stroke she remained comfortable and sharp of mind. In 2020 she was named cryptozoologist of the year by Loren Coleman and the International Cryptozoology Museum.

Marie Jeanne-Koffmann died July 11th 2021 , just eleven days shy of her 102nd birthday. She was a trail blazing and inspirational figure in the field . I can only hope her lifetime's collection of research notes in preserved for future cryptozoologists. I would have loved to have met her.

Marie Jeanne-Koffman, surgeon, mountaineer, cryptozoologist and almasty researcher, 19th of July 1919 - July 11th 2021 age 101. **RF**

EXTINCTION

EVERYONE GONE FOREVER

Search Extinction Rebellion / Follow us on social media for updates and events

Newsfile

Gloydius is a genus of venomous pitvipers endemic to Asia, also known as Asian moccasins or Asian ground pit vipers. Named after American herpetologist Howard K. Gloyd, this genus is very similar to the North American genus Agkistrodon. 22 species are currently recognized. A recent review of the molecular phylogeny of the genus by Jing-Song Shi, Jin-Cheng Liu, Rohit Giri, John Benjamin Owens, Vishal Santra, Sourish Kuttalam, Melvin Selvan, Ke-Ji Guo and Anita Malhotra has revealed two new species based on specimens collected from Zayu, Tibet,

west of the Nujiang River and Heishui, Sichuan, east of the Qinghai-Tibet Plateau.

The new species, *Gloydius lipipengi sp. nov.*, can be differentiated from its congeners by the combination of the following characters: the third supralabial not reaching the orbit (separated from it by a suborbital scale); wide, black-bordered greyish postorbital stripe extending from the posterior margin of the orbit (not separated by the postoculars, covering most of the anterior temporal scale) to the ventral surface of the neck; irregular black annular crossbands on the mid-body; 23-21-15 dorsal scales; 165 ventral scales, and 46 subcaudal scales. *Gloydius swild*

sp. nov. can be differentiated from its congeners by the narrower postorbital stripe (only half the width of the anterior temporal scale, the lower edge is approximately straight and bordered with white); a pair of arched stripes on the occiput; lateral body lakes black spots; a pair of round spots on the parietal scales; 21 rows of mid-body dorsal scales; zigzag dark brown stripes on the dorsum; 168–170 ventral scales, and 43–46 subcaudal scales.

The molecular phylogeny in this study supports the sister relationship between *G. lipipengi sp. nov.* and *G. rubromaculatus*, another recently described species from the Qinghai-Tibet Plateau, more than 500 km away, and indicate the basal position of *G. himalayanus* within the genus and relatively distant relationship to its congeners.

SOURCE: Jing-Song Shi, Jin-Cheng Liu, Rohit Giri, John Benjamin Owens, Vishal Santra, Sourish Kuttalam, Melvin Selvan, Ke-Ji Guo and Anita Malhotra. 2021. Molecular Phylogenetic Analysis of the Genus Gloydius (Squamata, Viperidae, Crotalinae), with Description of Two New Alpine Species from Qinghai-Tibet Plateau, China. ZooKeys. 1061: 87-108.

Liolaemus is a genus of iguanian lizards, containing many species, all of which

are endemic to South America. The diversity of reptiles in the Andes of southwestern Peru is poorly documented. Despite the fact that studies on saurians have intensified in recent years, mainly in the genus Liolaemus, information gaps on the biodiversity of this area remain. Such is the case of the Reserva Paisajística Subcuenca del Cotahuasi (RPSCC), Department of Arequipa, where populations of an undescribed species of the genus Liolaemus have been discovered recently. These individuals, now described as *Liolaemus warjantay*, have morphological and molecular characteristics that are not assignable to any of the known species.

SOURCE: Misshell D. Ubalde-Mamani, Roberto C. Gutiérrez, Juan C. Chaparro, Alvaro J. Aguilar-Kirigin, José Cerdeña, Wilson Huanca-Mamani, Stefanny Cárdenas-Ninasivincha, Ana Lazo-Rivera, and Cristian S. Abdala. 2021. A New Species of Liolaemus (Squamata: Liolaemidae) from the Reserva Paisajística Subcuenca del Cotahuasi, southwestern Peru. Amphibian & Reptile Conservation. 15(2) [Taxonomy Section]: 172–197 (e287).

New Zealand may not have any snakes, but there are many lizards including a diverse cool temperate assemblage of skinks, with 60+ identified taxa (genus Oligosoma Girard), of which only 50 have been formally described. A new species (*Oligosoma kakerakau sp. nov.*) has been described from Bream Head

Scenic Reserve, near Whangārei Heads, Northland. This species is considered to be conspecific with a single specimen (Oligosoma "Whirinaki") previously reported (in 2003) from Whirinaki Te Pua-a-Tāne Conservation Park ~370 km further south. Oligosoma kakerakau sp. nov. can be distinguished from all other members of the genus by a combination of a distinctive "teardrop" marking below the eye, a distinctive mid-lateral stripe, and the colouration and pattern on its ventral surface.

The phylogenetic analyses indicate that *Oligosoma kakerakau sp. nov.* is most closely related to *O. zelandicum* (Gray), and more distantly to *O. striatum* (Buller) and *O. homalonotum* (Boulenger). Sea level changes during the Pliocene, such as the formation of the Manawatū Strait, may have contributed to the divergence between *Oligosoma kakerakau sp. nov.* and *O. zelandicum.*

BSOURCE: Benjamin P. Barr, David G. Chapple, Rodney A. Hitchmough, Geoff B. Patterson and Ngāti-wai Trust Board. 2021. A New Species of Oligosoma (Squamata: Scincidae) from the northern North Island, New Zealand. Zootaxa. 5047(4); 401-415.

Another new species of Oligosoma has also been described from a slate scree in montane tussock grassland in Kahurangi National Park, New Zealand, where it is currently known from a single small site.

The new species (*Oligosoma kahurangi sp. nov.*) can be distinguished from all congeners by its extremely long tail, 36–38 mid-body scale rows, head length/head width ratio of 1.66, and colour pattern. It is part of the *O. longipes Patterson* species complex.

The species is currently very poorly known but likely to be highly threatened, and it has been suggested that it should be listed as Nationally Critical (Data Poor, One Location) in New Zealand, and Data Deficient in the IUCN red-list. Predation by introduced mammals, particularly mice, is assumed to be a threat to its survival.

SOURCE: Geoff B. Patterson and Rodney A. Hitch-mough. 2021. A New Alpine Skink Species (Scincidae: Eugongylinae: Oligosoma) from Kahurangi National Park, New Zealand. Zootaxa. 4920(4); 495–508.

narrow strip of the Pacific Coast, where it is entirely surrounded by B. nigriventris, a distant relative. Although intraspecific molecular variation is almost entirely absent, some population structure was detected across the 4 km extent of its range. Because of its tiny range and limited genetic variation, the impacts of any potential modifications to its known habitat should be evaluated to ensure the species' continued conservation.

SOURCE: Samuel S. Sweet and Elizabeth L. Jockusch. 2021. A New Relict Species of Slender Salamander (Plethodontidae: Batrachoseps) with a Tiny Range from Point Arguello, California. Ichthyology & Herpetology, 109(3); 836-850.

Among western North American amphibian lineages, the plethodontid salamander genus Batrachoseps has undergone the most extensive radiation. A new species - Batrachoseps wakei - in the genus from the vicinity of Point Arguello, central California has been described. This lineage falls within the B. pacificus group, but it is differentiated from other species in the group by both molecular sequence data and morphology.

Something particularly interesting is its biogeography. It is geographically disjunct from its close relatives, with a tiny range in unlikely habitat along a

Fourteen species of bats in the genus Molossus currently are recognized in the

Neotropical region; only three are known from Argentina. However a new species has been described based on specimens collected in the province of Santa Fe, Argentina, in the Pampa ecoregion. The new species - Molossus melini - can be distinguished from its congeners by its general strong orange coloration, forearm length > 41 mm, dorsal hairs bicolour and long (~5 mm), infraorbital foramen laterally oriented, and long and forward-projected (pincer-like) upper incisors.

SOURCE: M. Eugenia Montani, Ivanna H. Tomasco, Ignacio M. Barberis, Marcelo C. Romano, Rubén M. Barquez and M. Mónica Díaz. 2021. A New Species of Molossus (Chiroptera: Molossidae) from Argentina. Journal of Mammalogy. Gyab078

Thirteen species of West Indian boas (Chilabothrus) are distributed across the islands of the Greater Antilles and Lucayan Archipelago. Hispaniola is unique among this group of islands in having more than two species of Chilabothrus — three are currently recognized. However, a fourth species from Hispaniola has now been described. Chilabothrus ampelophis is a newly discovered distinctive species of small boa from the dry forest of the Barahona Peninsula, southwestern Dominican Republic, near the border with Haiti.

This new species resembles in body size and in other aspects its closest relative Chilabothrus fordii (Günther 1861), with which it appears to be allopatric. The new species, which we describe as Chilabothrus ampelophis sp. nov., differs from C. fordii in body, head, and snout shape; in scalation; in both coloration and colour pattern; and in

phylogenetic uniqueness. The discovery of this new species is especially important as it appears to be among the smallest boid (Boidae) species, has an arboreal specialization, and is found in a very restricted and highly threatened habitat.

SOURCE: Miguel A. Landestoy T., R. Graham Reynolds and Robert W. Henderson. 2021. A Small New Arboreal Species of west Indian Boa (Boidae; Chilabothrus) from southern Hispaniola. Breviora. 571 (1):1-20.

CFZ investigators, Carl Marshall and Andrew Jackson, visited the central American country of Belize in 2009 in search of information about various Central American cryptids including the sisemite and the lusca.

While they were there they visited some small cays off of the Belizian coast, off Placencia, and were surprised to find that nearly every tree contained an osprey nest, and - further down the trunk - a distinctive small boa.

These snakes have been described as distinctive subspecies of Boa constrictor, and some have even named it as *Tropidophis greenwayi* a distinct species.

Evidence such as this of the newly discovered dwarf boa from Hispaniola suggests that speciation in isolated, and even not so isolated, populations of boid is a reasonably common occurrence. Therefore it could well be that some, if not all, of these island populations discovered by Marshall and Jackson are distinct subspecies if not distinct species.

Barbodes pyrpholeos, is the first cave-dwelling cyprinid fish reported from the

Philippines. It is described from karst systems in Mindanao. It is distinguished from other congeners by having a poorly pigmented body with reddish fins in combination with a smooth dorsal-fin spine without serrations, and several additional morphological characters. It differs from the Indonesian troglobitic congener *Barbodes microps* by the presence of eyes, and a narrower body amongst other characters. *Barbodes montanoi*, a putative close relative of the new species, is redescribed based on recently collected material.

SOURCE: Tan Heok Hui and Daniel Edison M. Husana. 2021. Barbodes pyrpholeos, New Species, the First Cave-dwelling Cyprinid fish in the Philippines, with Redescription of B. montanoi (Teleostei: Cyprinidae). RAFFLES BULLETIN OF ZOOLOGY. 69; 309–323.

A new species of Enyalioides - *Enyalioides feiruzae* - has been discovered from the premontane forest of the Río Huallaga basin in central Peru.

The most similar and phylogenetically related species are *E. binzayedi* and *E. rudolfarndti*.

However, the new species differs from *E. binzayedi* (state of character in parentheses) by having dorsal scales strongly keeled on paravertebral region and feebly keeled or smooth elsewhere (prominent medial keel on each dorsal scale), more dorsals in transverse row between dorsolateral crests at midbody 26–39, x̄ = 30.44 (22–31, x̄ = 27.57), and a conspicuous posteromedial black patch in the gular region of males (absent).

Contrarily, adult males of the new species and *E. rudolfarndti* are readily distinguished by having a conspicuous posteromedial black patch in the gular region, absent in *E.*

rudolfarndti, and by lacking a conspicuous orange blotch (faint if present) on the ante humeral region, as in *E. rudolfarndti*.

SOURCE: Pablo J. Venegas, Germán Chávez, Luis A. García-Ayachi, Vilma Duran and Omar Torres-Carvajal. 2021. A New Species of Wood Lizard (Hoplocercinae, Enyalioides) from the Río Huallaga Basin in Central Peru. Evolutionary Systematics. 5(2): 263-273.

A distinct new species of the genus Hemiphyllodactylus has been discovered. *Hemiphyllodactylus goaensis sp. nov,* has been described based on three specimens collected from semi-urban areas in Goa state of India. The new species can be easily distinguished from all peninsular Indian congeners by its small body size (SVL up to 32.4 mm), having 16–18 dorsal scales and 13 or 14 ventral scales at mid-body contained within one longitudinal eye diameter, nine or ten precloacal pores separated by 1–5 poreless scales from a series of 10–12 femoral pores on each thigh in males, lamellar formula of manus 2222 and of pes 2323 & 2333, as well as subtle colour pattern differences. Mitochondrial sequence divergence confirms the distinctiveness of the new species, which is not closely allied to either the South Indian or Eastern Ghats clades of Indian Hemiphyllodactylus and appears to be a member of a third Indian Hemiphyllodactylus clade.

Hemiphyllodactylus goaensis sp. nov. is the first member of the genus to be described from the northern Western Ghats region as well as Goa state, and also only the second Indian Hemiphyllodactylus known from < 100 m asl. *Hemiphyllodactylus goaensis sp. nov.* extends the known distribution of the genus in western India ~ 560 km north in aerial distance and highlights that the genus is more widely distributed than previously thought and most likely contains numerous undescribed species. We also provide final museum numbers for type specimens of *H. arakuensis* and the holotype of *H. kolliensis*.

SOURCE: Akshay Khandekar, Dikansh S. Parmar, Nitin Sawant and Ishan Agarwal. 2021. A New Species of the Genus Hemiphyllodactylus Bleeker, 1860 (Squamata: Gekkonidae) from Goa, India. Zootaxa. 5027(2); 254-268.

The hill stream loach genus Indoreonectes is endemic to peninsular India south of the Satpura hill ranges and is represented by three species *I. evezardi, I. keralensis* and *I. telanganaensis*. *Indoreonectes evezardi* has been suggested as a species complex based on recent genetic studies; however, due to lack of type material the species delimitation has been difficult.

I. evezardi collected from its type locality has been re-examined and two new species have been described from

Indoreonectes evezardi (Day, 1872),

Indoreonectes neeleshi &
Indoreonectes rajeevi

the northern Western Ghats of India. *Indoreonectes neeleshi*, described from Mula River tributary of Godavari river system, can be diagnosed from all its congeners based on a combination of characters: inner rostral barbel reaching middle of nostril; maxillary barbel reaching midway between eye and posterior border of operculum; dorsal hump behind nape; bars on lateral side of the body wider than inter-bar space; total vertebrae 35 and dorsal fin insertion between 13th and 14th abdominal vertebrae.

Indoreonectes rajeevi, described from Hiranyakeshi River of the Krishna river system, differs from all its congeners based on a combination of characters: inner rostral barbel reaching anterior margin of eye; maxillary barbel reaching posterior border of operculum; conspicuous black markings on lower lip, dorsal hump absent; total vertebrae 36 and dorsal fin insertion between 12th and 13th abdominal vertebrae.

SOURCE: Pradeep Kumkar, Manoj Pise, Pankaj A. Gorule, Chandani R. Verma and Lukáš Kalous. 2021. Two New Species of the Hillstream Loach Genus Indoreonectes from the northern Western Ghats of India (Teleostei: Nemacheilidae). Vertebrate Zoology. 71: 517-533.

A new species of giant guitarfish, *Glaucostegus younholeei sp. nov.,* has been described from 13 specimens, 730–933 mm total length, collected from fish landing center of Bangladesh Fisheries Development Corporation in Cox's Bazar district of Bangladesh. The new species is distinguished from congeners in having the following combination of characters: Body brownish or greyish in colour with a narrowly wedge-shaped disc, and long narrow bluntly pointed snout (angle 31–40°), and broad oblique nostrils with the narrow anterior opening. Nostrils about half of the mouth width, subequal (0.98–1.33) to

381 mm

internasal width; ~55–57 nasal lamellae; anterior nasal flaps slightly penetrating into internasal space, their interspace 2.20– 2.61 in length of the posterior nasal aperture. Orbit very small in adults, diameter 8.19–11.62 in preorbital length, 2.25–2.69 in interorbital space. Rostral ridges almost joined along their entire length; margin of cranium sharply demarcated before eyes. Spiracular folds very short and widely separated. Skin rough, densely covered with small denticles, more coarsely granular on the dorsal surface than ventrally, enlarged between orbits and in a distinct band between nape and first dorsal fin. Tail relatively longer, length 1.15–1.48 in disc length; dorsal fins narrowly spaced, interspace 1.32–2.11 in base length of the first dorsal fin. Clasper length in adult male 4.37–5.70 in total length. Phylogenetic analysis of DNA barcode sequences also shows the clear divergence of *Glaucostegus younholeei* from other congeneric species obtained from GenBank.

SOURCE: Kazi Ahsan Habib and Md Jayedul Islam. 2021. Description of A New Species of Giant Guitarfish, Glaucostegus younholeei sp. nov. (Rhinopristiformes: Glaucostegidae) from the northern Bay of Bengal, Bangladesh. Zootaxa. 4995(1); 129-146.

Calamaria dominici sp. nov. a new species of Calamaria Boie, 1827 has been described based on a single specimen collected in evergreen forest at 1240 m elevation of Ta Dung Nature Reserve in Dak Nong Province, Central Vietnam.

The new species is characterized by (1) rostral wider than high; (2) paraparietal surrounded by six shields and scales; (3) eye diameter larger than eye mouth distance; (4) preocular present; (5) supralabials 5/4, 3-4/2-3 entering orbit; (6) maxillary teeth nine, modified; (7) infralabials 5/4, first three touching anterior chin shields; (8) mental touching tip of right anterior chin shield; (9) ventrals 1 + 174; subcaudal scales 18/17, divided; (10) precloacal plate single; (11) tail relatively short (6.2% of the total length), nearly as thick as body, slightly tapering, and ending in obtuse point; (12) dorsal scales reducing to six rows at position above 4th subcaudal, and to four rows above 13th subcaudal on tail; (13) dorsum dark with irregular yellow blotches; and (14) ventral side dark with few yellow blotches and bands. This is the sixth new Calamaria species described from Vietnam in the past thirteen years and the tenth species of Calamaria recorded from this country.

SOURCE: Thomas Ziegler, Vu A. Tran, Randall D. Babb, Thomas R. Jones, Paul E. Moler, Robert W. Van Devender and Truong Q. Nguyen. 2019. A New Species of Reed Snake, Calamaria Boie, 1827 from the Central Highlands of Vietnam (Squamata: Colubridae). Revue suisse de zoologie; annales de la Société zoologique suisse et du Muséum d'histoire naturelle de Genève. 126 (1); 17-26.

Proceratophrys korekore is a New Species of Proceratophrys Miranda-Ribeiro, 1920 (Anura, Odontophrynidae) from Southern Amazonia, Brazil Based on concordant differences in morphology, male advertisement call, and 16S mtDNA barcode distance, a new species of Proceratophrys has been described from southern Amazonia, in the states of Mato Grosso and Pará, Brazil. The new species is most similar to *P. concavitympanum* and *P. ararype* but differs from these species by its

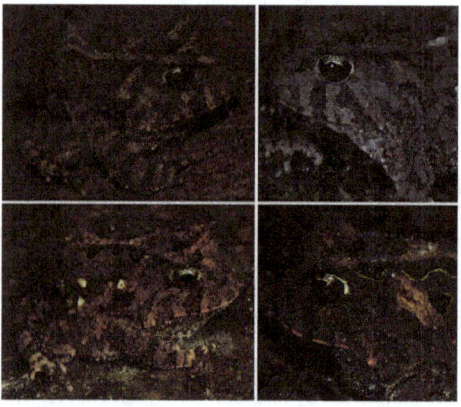

proportionally larger eyes and features of the advertisement call.

Using an integrative approach (molecular, bioacoustics, and adult

morphology), the authors were able to distinguish the new species from other congeneric species. The new species is known only from the type locality where it is threatened by illegal logging and gold mining as well as hydroelectric dams.

SOURCE: Diego J. Santana, Leandro Alves da Silva, Anathielle Caroline Sant'Anna, Donald B. Shepard and Sarah Mângia. 2021. A New Species of Proceratophrys Miranda-Ribeiro, 1920 (Anura, Odontophrynidae) from Southern Amazonia, Brazil. PeerJ. 9:e12012 .

A new species of Andinobates (Dendrobatidae) has been described from the East Andes of Colombia, just 37 km away from the Colombian capital, Bogotá. *Andinobates supata sp. nov.*, represents the second known species of yellow Andinobates, and can be distinguished from the other, *Andinobates tolimensis*, by an unique combination of ventral and dorsal color patterns.

A BUNCH OF BATRACHIANS

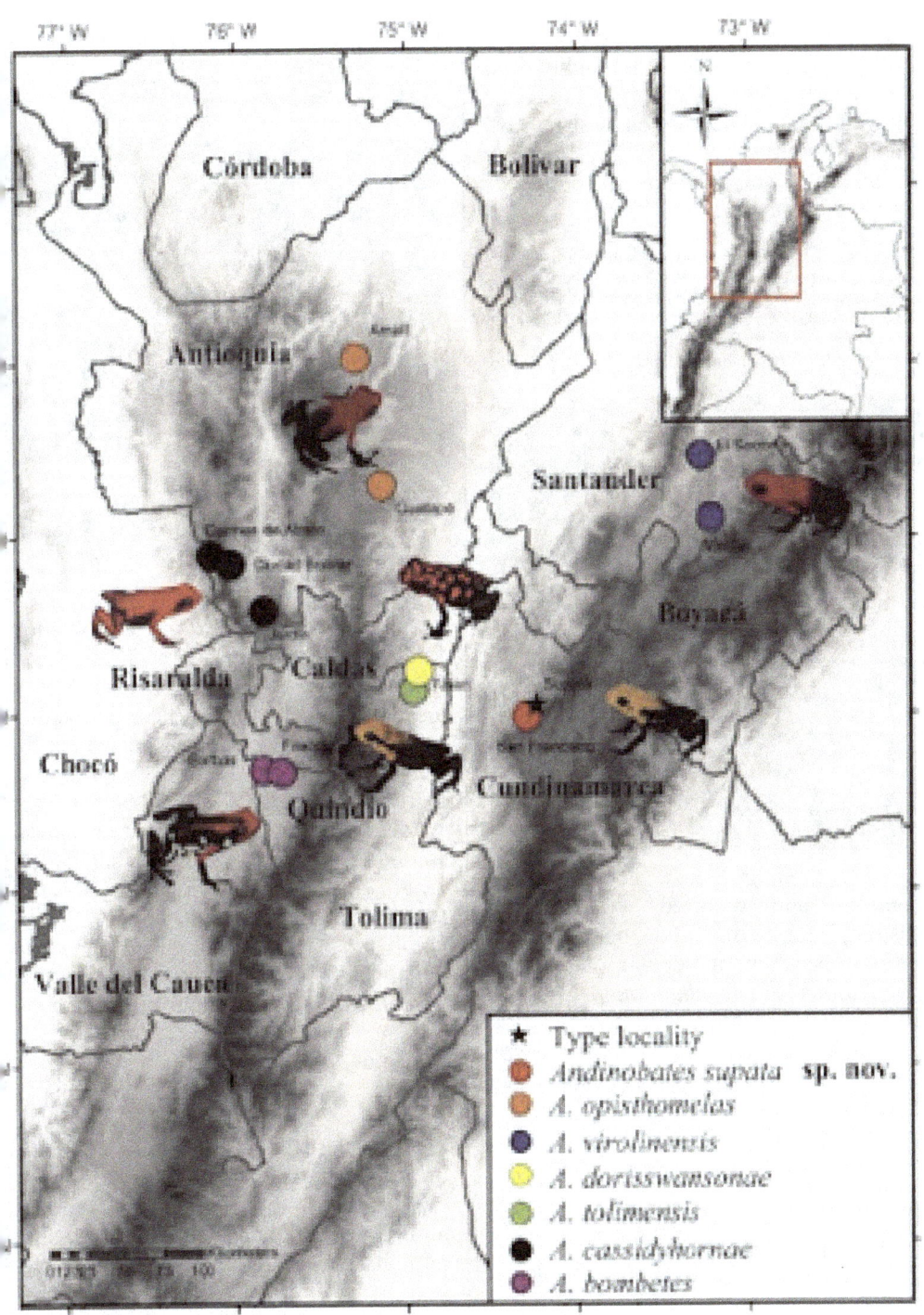

77° W   76° W   75° W   74° W   73° W

Córdoba   Bolívar

Antioquia

Santander

Risaralda   Caldas

Boyacá

Chocó

Quindío   Cundinamarca

Tolima

Valle del Cauca

★ Type locality
● *Andinobates supata* **sp. nov.**
● *A. opisthomelas*
● *A. virolinensis*
● *A. dorisswansonae*
● *A. tolimensis*
● *A. cassidyhornae*
● *A. bombetes*

The phylogenetic analyses, based on ≈ 1120 bp from two mitochondrial markers (16S rRNA and cytochrome b) showed that this new taxon is sister to a clade formed by *A. cassidyhornae, A. bombetes, A. opisthomelas, A. tolimensis and A. virolinensis.*

Giovanni A. Cháves-Portilla, Erika Nathalia Salazar, José Gil-Acero, Adriana Dorado-Correa, Roberto Márquez, José Vicente Rueda-Almonacid and Adolfo Amézquita. 2021. A New Species of Andean Golden Poison Frog (Andinobates, Dendrobatidae) from the Eastern Andes of Colombia. Zootaxa. 5047(5); 531-546.

Amolops chunganensis is a species complex and reported widely from eastern, southern, and southwestern China. Based on molecular data of 19 populations of *A. chunganensis sensu lato* from China, including the population from Mt. Wuyi (type locality), a new species has been described *Amolops chaochin sp. nov.*, from southwestern China, which was previously identified as *A. chunganensis.*

SOURCE: Ke Jiang, Jin-Long Ren, Zhi-Tong Lyu, Dan Wang, Zeng Wang, Ke Lv, Jia-Wei Wu and Jia-Tang Li. 2021. Taxonomic Revision of Amolops chunganensis (Pope, 1929) (Amphibia: Anura) and Description of A New Species from southwestern China, with Discussion on *Amolops monticola* group and Assignment of Species Groups of the Genus Amolops. Zoological Research. 42(5); 574-591.

A new species of bromeliad-dwelling Pristimantis treefrog has been described from primary montane forest (2,225 m a.s.l.) in southern Peru. The type locality is near Thiuni, in the Department of Puno (province of Carabaya) in the upper watershed of a tributary of the Inambari River. *Pristimantis achupalla sp. n.* is characterized by a snout-vent length of 10.0–12.8 mm in adult males (n = 4), unknown in adult females, and is compared morphologically and genetically with species in the *Pristimantis lacrimosus* group.

The new species is characterized by having skin on dorsum and flanks rugose, green brownish colour, distinctive scapular folds, subacuminate or acuminate snout profile, upper eyelid bearing two or three subconical tubercles and some rounded tubercles, rostral papilla, flanks light brown to brown, with irregular dark brown marks.

SOURCE: Alex Ttito and Alessandro Catenazzi. 2021. *Pristimantis achupalla* sp. n., A New Minute Species of Direct-developing Frog (Amphibia, Anura, Strabomantidae) inhabiting Bromeliads of the Montane Forest of the Amazonian Andes of Puno, Peru. PeerJ. 9:e11878.

Nurse frogs of the genus Allobates are mostly small with cryptic colouration, with the exception of the *Allobates femoralis* group that has bright colours. They are mostly terrestrial creatures, found in the leaf litter of tropical rain forests. Most species deposit eggs in the leaf litter; tadpoles are transported to the water on the backs of the parents. *Allobates nidicola* and *Allobates chalcopis*, however, have endotrophic tadpoles that develop into froglets in the nest, without entering water.

A new species of nurse-frog (Aromobatidae, Allobates) from the Amazonian forest of Loreto, Peru has been described using morphological, acoustic and genetic data. Our phylogenetic analysis placed *Allobates sieggreenae sp. nov.* as the sister species of *A. trilineatus*, the most similar-looking species and with which it was previously confused. However, the new species has a brown dorsum, solid dark brown lateral dark stripe not fading towards groin, adult males with few and sparse melanophores over a cream background on chin, chest, and belly, dark transverse bars absent on thighs, and an advertisement call formed by a trill of single notes (in *A. trilineatus dorsum* dark brown, blackish brown lateral dark stripe, paler from mid-body to groin, adult males with a dark background colour on chin, chest, and belly due to a dense layer of melanophores, dark transverse bar present on dorsal surface of thighs, and trills of paired notes). *Allobates sieggreenae* is known from two localities of Amazonian white-sand forest ecosystems east of the Ucayali River.

SOURCE: Giussepe Gagliardi-Urrutia, Santiago Castroviejo-Fisher, Fernando J. M. Rojas-Runjaic, Andrés F. Jaramillo, Samantha Solís and Pedro Ivo Simões. 2021. A New Species of Nurse-Frog (Aromobatidae, Allobates) from the Amazonian Forest of Loreto, Peru. Zootaxa. 5026(3); 375-404.

Just in case you have been wondering why we always keep the newly discovered batrachian species separate, ad this is because there are so many of them. It is hard not to see them as parts of Mother Nature's system of checks and balances to compensate for all the batrachian species that are being flushed into extinction by the chytrid virus.

You may think that this is a ridiculous point of view, but having ridiculous points of view is what being Forteans is all about.

Man Beasts (BHM)

Over the past few years quite a lot of images and video of North American black bears is walking bipedally have circulated across the Internet. Many of these are images of bears who have somehow injured, or even lost, their front paws, and as a result, have no option, but to walk on their two hind legs. And it has been suggested by many people that these bipedal ursids are responsible for some, if not all, bigfoot reports, and whilst we would not go so far as to say that there are no mystery higher primates in the places where BHM reports have been made, we would suggest that it would be foolhardy to dismiss the suggestion that many of these reports are due to upright-walking bears out of hand.

We were very interested, therefore, recently to read a fascinating piece of research by Floe Foxon. Over the years, various scientists have suggested that American black bears have been

responsible for some of all Bigfoot sightings. However as the ScienceAlert.com website wrote on the 31st January:

"Foxon has now expanded on prior results by extending the analysis to all places in the US and Canada where black bears and humans live near one another.

The data he used for Bigfoot sightings came from the Bigfoot Field Researchers Organization, which keeps a geographic database of eyewitness reports mostly from the twentieth century onwards.

Foxon then compared this information to local data on black bear density and spread as well as human population densities. He says this an improvement on the simplified projections used in previous papers.

According to Foxon's rigorous regression model – which shows if changes seen in one variable are associated with changes in another –

Bigfoot sightings are largely explained by misidentified black bears."

We found his paper very interesting and to our embarrassment we had never heard of "regression analysis". Indeed, to our even greater embarrassment, when I first read they had used "regression analysis" to analyse a large number of "sasquatch" sightings, we initially thought that this was going to be something akin to the strange folk who hypnotise people in order to regress them to a so-called "past life". It might have been amusing if it had actually been so, but we think that what they actually have done is of enormous importance.

According to those jolly nice folk at Wikipedia:

"In statistical modelling, regression analysis is a set of statistical processes for estimating the relationships between a dependent variable (often called the 'outcome' or 'response' variable, or a 'label' in machine learning parlance) and one or more independent variables (often called

'predictors', 'covariates', 'explanatory variables' or 'features'). The most common form of regression analysis is linear regression, in which one finds the line (or a more complex linear combination) that most closely fits the data according to a specific mathematical criterion. For example, the method of ordinary least squares computes the unique line (or hyperplane) that minimizes the sum of squared differences between the true data and that line (or hyperplane). For specific mathematical reasons (see linear regression), this allows the researcher to estimate the conditional expectation (or population average value) of the dependent variable when the independent variables take on a given set of values. Less common forms of regression use slightly different procedures to estimate alternative location parameters (e.g., quantile regression or Necessary Condition Analysis) or estimate the conditional expectation across a broader collection of non-linear models (e.g., nonparametric regression).

We think that we know what all this means, although we will be the 1st to admit that most of what we know about the science of statistical analysis comes from the pages of various books by Isaac Asimov, when he discusses his hypothetical science of psychohistory, which seems to be a mirror image of what Foxon has done here

Mystery Cats

Spanish bombs in Andalucía

In the autumn of 2020, this photograph was published in an English language newspaper targeted at expats. It reopened the debate over whether there are panthers wild in Spain's Andalucía.

https://www.theolivepress.es/spain-news/2020/09/16/new-picture-of-black-panther-re-ignites-search-in-sleepy-village-in-spains-andalucia/?

MYSTERY CATS STUDY GROUP

fbclid=IwAR162KKUQSHHg6r78Q-5TeWCepDijA4EOt-qpHOHnjmhqH1zhkcDgcgJZ3E

Despite the fact that it has been claimed that the panther is the escaped pet of an unnamed gangster, the photo published in The Olive Press article – unfortunately doesn't show anything more than a large domestic cat. However, we should bear in mind for future reference that Ventas de Huelma, where the picture was taken, and where there have been further reports, is approximately only 100 miles south-east (as the crow flies, or as the cat meanders!) of the nearest Iberian lynx sub-population (Lynx pardinus) near Cordoba; and even though [this time] we can be almost certain that the cat photographed is a just a very robust domestic of the species F. catus (melanistic individuals can be larger

than normal due to the influence of a pleiotropic gene triggering two or more seemingly unrelated phenotypic traits, which, in this case, can cause not only black pelage, but also a larger than average body-size), it is not beyond the realm of possibility that some of the remaining Iberian lynxes could on occasion venture this far south in search of suitable habitats.

But Iberian lynx aren't black I hear you cry! Genetic drift (limited gene flow) can, and does, cause recessive alleles to be recurrently expressed, and the non-agouti allele, which is apparently the gene which triggers melanism in the bobcat (Lynx rufus - the only member of the Lynx genus known to occasionally display melanism), is of course recessive. It might therefore be theoretically possible that other rare lynx populations, other than the already mentioned L. rufus, might also, albeit very rarely, produce this mutation unnoticed. By the turn of the 21st century, the Iberian lynx was on the verge of extinction, as only about 100 individuals survived in two isolated sub-populations in Andalusia. Conservation measures have been implemented since 2002 which included improving habitat, restocking of rabbits, translocating, re-introducing and monitoring Iberian lynx. By 2012, the population had increased to 326 individuals and is now somewhere around 400.

Lynx pardinus is therefore one of the world's rarest felids, and one still considered to be surviving in a genetic "bottleneck" situation. If any lynx species other than the bobcat is likely to start producing random black variants, L. pardinus, in my opinion, is probably going to be it!

This is all, however, irrelevant this time, as the cat photographed above is non other than a big (meaning a robust) black domestic cat. That being said, if this mutation proved beneficial in countries with little or no large apex predators, whether naturally like New Zealand, or because of the desertification of the landscape/prolonged hunting, like Australia, and dare I say, Great Britain, might these large domestics begin to fill the vacant ecological niches once occupied by species now absent from the country's respective fauna – it's free for the taking, and evolution is a dynamic process!

This is obviously entirely subjective, but in my opinion at least, it's a fascinating thought to bear in mind. This would also explain many of the blurry photographs we get each year of domestic cats where the observer has been adamant that what he or she observed was much larger than the average F. catus.

It might make up a percentage of ABC reports, at least. CM

A Helensburgh Happening

This photo was allegedly taken near Helensburgh in Scotland. A Panther said the press. Carl Marshall disagreed.

**https://
www.helensburghadvertiser.co.uk/
news/18446909.big-cat-loose-
helensburgh/**

Arguably, the most famous alleged case of a big cat from Helensburgh (Argyll and Bute, Scotland) occurred in July 2009, when MoD Police dog handler Chris Swallow, reported seeing a large black cat close to the West Highland railway line, near his home.

Unfortunately it was nothing of the sort!

But, as is often the case with ABC reports, once a story makes headlines, that's it, it become set in stone, especially when the witness is obviously credible and very unlikely to have created a hoax. However, what Mr Swallow observed was, without doubt, a black domestic cat. It just so happened that from his observation point, which was positioned in such a way that made the tracks seem considerably closer together than they actually are; thus making the cat in question look larger.

This can be conclusively proven if one watches the entire unedited video, which, as the frame pans out, more of the track farther along the line is revealed, which then veers round to the right, lining up perfectly now with the observer and showing the actual width of the track, and correcting the illusion. Unfortunately the cat doesn't continue along the line to this point. Mr Swallow, by focusing on framing the animal with his camera, would likely have not noticed this at the time. There is unfortunately little unanimity among ABC researchers. And we must do a better job of separating the 'wheat from the chaff' if we are ever going to make better sense of the British phenomenon. The recent case published on May 9th seems genuine enough, in the sense the location of the sighting (Helensburgh Golf Club's course) is the kind of place where wildlife and man do come into contact, and during the covid 19 lockdown, who is to say these animals haven't been making the most of our absence, in fact, they likely have!

Here we have two eyewitnesses, a man and his wife (both unnamed), observing an animal at a distance close enough to be identifiable. With that being said, one must also consider whether the couple had been influenced by the erroneous, but nonetheless well known, 2009 report.

There have been other reports made from Helensburgh, other than the two examples mentioned here; even a few observations of faun coloured big cats believed to be pumas. So who knows. Unfortunately, again, we don't have much to work with. We must therefore remain cautious, but not overcautious.

Our friends at Dragonfly Films should be congratulated for the latest in their series of British Bigcat documentaries. Check it out at Vimeo at the following:

https://vimeo.com/ondemand/pantherabritannia

Aquatic Monsters

Loch Ness blobs

According to the *Daily Mirror*, "the existence of the Loch Ness monster has been proven according to a veteran Nessie hunter who spotted a mysterious presence in the water."

This claim was made by Eoin O'Faodhagain, aged 58, who said he got the shock of his life when he saw a strange black shape in the loch, estimated to be 10ft long.

Eoin O'Faodhagain, is by most metrics a veteran Loch Ness monster investigator, although his highest profile results have all been from the Loch Ness webcams and although we don't want to appear to be negative about the whole affair, until this latest picture/video, the pictures he has taken from the webcam are all morpheus blobs in the middle distance which - truthfully - could be anything.

The new footage is considerably better, but we cannot say that it is definitely of "the monster" let alone conclusive proof of anything. It does appear to show a long object, low in the water, moving along quite determinedly, but that object could

be a log of some kind. On all our visits to Loch Ness we have discovered that there is something peculiar in the formation of the various currents within the lake, and as a result, there is a whole plethora of strange waveforms produced. We are not prepared to say whether these images produced by O'Faodhagain are not the result of one of these peculiar waveforms.

We are sure that as there are people within the cryptozoology community who believe all sorts of arcane things, there is somewhere a large collection of people who follow O'Faodhagain devotedly. Therefore, we are quite prepared for the torrent of abuse which we are going to be receiving once these aforementioned people read this section of the magazine. However, we have to tell it as we see it.

Something else which is incredibly important is that we are not trying to belittle

Mr O'Faodhagain in any way, shape or form. We are sure, from what we have read about him, that he is a very sincere man who has dedicated hundreds, if not thousands, of hours of his time to scrutinise the Loch Ness webcams. We are certain that each of the pictures that he has captured, were produced completely in good faith. Moreover, we would not wish to be in the position where we are discouraging people like Mr O'Faodhagain doing what they do. Indeed, it is very important and we sincerely hope it will eventually bear fruit and that the Loch Ness webcams will one day produce a photograph of a recognised giant creature.

Further than that, we sincerely hope that when this day does happen, (and we think it will), we hope that it is Eoin O'Faodhagain who takes the photography that will eventually prove the mystery.

Other Ogopogo

Now, to go to the other side of the world, more specifically to Lake Okanagan in British Columbia. On the 19th October last year, Global News presented the latest twist in the ongoing saga of Canada's best known lake monster.

https://globalnews.ca/video/9211789/possible-ogopogo-sighting-on-okanagan-lake

Dale and Colleen Hanchar along with their friend Myrna Brown, were sailing on Lake Okanagan when they saw a peculiar thing in the water. Dale told Global News:

"As a boater I was just checking if this was something dangerous which needed to be marked, so that somebody doesn't run into it, like a dead head or something similar. We went on by, and got thinking about it. I said to Myrna and my wife that it didn't look right and we've got to go look at that again". They managed to get their sailing boat within ten feet of the object and took a single photograph.

"We were all puzzled as to what that could possibly be. You know, we kind of eliminated in our heads what it wasn't. We talked about it a little bit and then we just kept on going."

However, the photograph was taken on a mobile phone and they were not able to decipher much through the small screen of the device. But, when they got home they looked at it on a big screen and were even more mystified. Somebody suggested that this could be an image of the elusive lake monster but when the picture was dispersed across the internet, the reaction was mixed. Some, like me, thought that it was an image of a rubber wolf-head mask floating in the water. Others such as cryptid researcher, folklorist and Fortean, Adam Benedict told Global News:

"What caught my attention was the two protruding objects at the surface. But when I zoom in on a larger screen, I see a water bird of some sort in the process of a dive, either just right below the surface or in the process of coming up. The two protruding objects at the back whether they are bent or they are kicking, but you can clearly see just below the surface an arm as well at its beak right at the top of the water line."

Benedict's theory was met with derision

from certain quarters of the cryptozoological community.

CFZ's very own Ian Squibbs, however, pointed out that it was only just over four weeks previously that the Kelowna Dragon Boat Club had held their annual dragon boat festival. He suggests, and it is hard not to disagree with him, that since many of the dragon boats are decorated by hand by their owners, the creature photographed in the lake was the unlucky remnant of one of the boats whose dragon head had detached from the rest of the craft.

Krakens in Spain: from myth to reality. By Javier Resines

Last June, the first recording of a giant squid in Spanish waters was captured, only the second - worldwide - after another filmed by a Japanese team in 2013. This made us turn our gaze to the unknown vastness of the oceans, the last corner of our planet to explore, the place where the impossible can become reality. The kraken not long ago adorned the boundaries of medieval maps, warning intrepid sailors of what they could find at the ends of the known world. As a kind of messenger of Evil, a sea creature from Norse legends. The myth tells that, due to its size, ships could confuse it with an island on which - on occasion - they landed, only to sink deep in the sea when the monster decided to attack. In fact, it does not seem that such a being has ever existed. Surely, exaggeration is behind many of the horrifying anecdotes that we have been told. But what does exist today, although Science has denied it until a few

GIANT SQUID - FIFTY-FIVE FEET LONG.

decades ago, is a group of species that seem to embody the old fears of encounters with ferocious marine beings: the Giant Squid.

Myth becomes reality

But what do we mean when we talk about large cephalopods? Basically, two species that - fortunately for the researchers - are abundant in Spanish waters: taningia danae and, above all, architeuthis dux.

Architeuthis is a genus of cephalopods that can reach up to 22 meters in length and almost 300 kg in weight, although it is speculated that adult specimens can weigh more than a ton and reach even larger sizes. However, to date, only young individuals have been captured.

They usually live between 400 and 1500 meters deep and seem to be an animal with solitary habits. They have almost never been seen in pairs or, even less frequently, in banks or other groups. Their reproductive cycle is unknown, but we do know that they contain immense amounts of ammonia to maintain their buoyancy, which makes it inedible for man, but not for sperm whales, pilot whales and the occasional predator.

They have the largest eyes in the animal kingdom (up to 25 centimetres in diameter) and are the being with the fastest growth rate known to Science, between 3 and 5 centimetres a day.

The other species, taningia danae, are a less common animal than the previous one (which are already difficult to find) and for which we have even less information. They are a shorter and wider squid than architeuthis and live between 500 and 1000 meters deep and, instead of

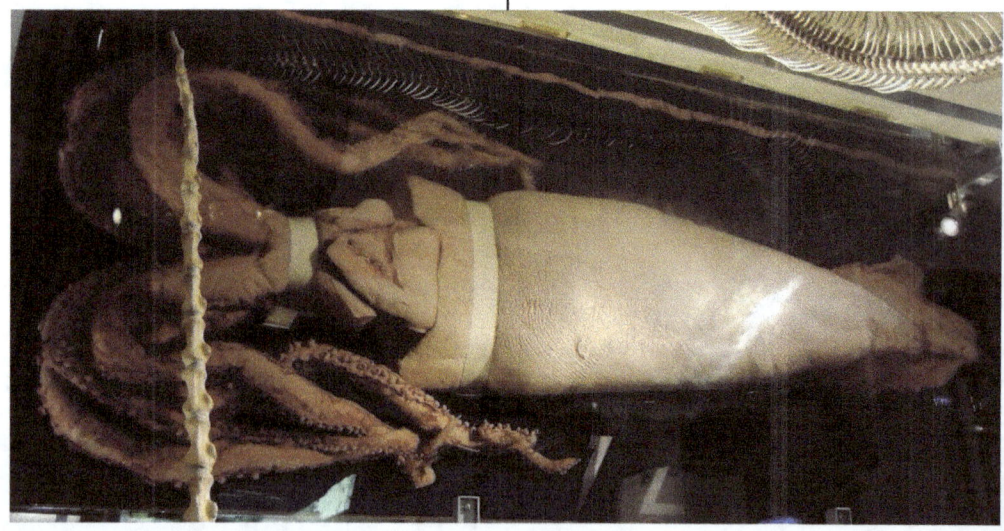

suction cups, they have two rows of small claws or hooks with which it captures its prey. The most extraordinary thing about Taningia are the two luminous organs they possess - called photophores - are capable of emitting a peculiar yellow light with which it seems to attract its prey and - probably - maintain basic communication with other individuals of their species.

Both species cohabit in the existing pools off the Asturian coast and also in some areas of the Canary Islands. Spain is a privileged country in terms of the quantity and quality of the encounters that have occurred with giant squid. In our waters we have detected four main areas in which our gigantic friends are present: the Canary Islands, Asturias, the Strait and Levante area and also off the Galician coast.

Canary Islands, paradise of the giant squid

Between the south of Tenerife and the east of La Gomera there is a corridor that houses a colony of giant squid. Sightings of these animals have also occurred in Fuerteventura, south of Gran Canaria and east of La Palma. Since ancient times, countless remains of tentacles have been seen floating in Canary waters and, up to thirty times, an entire body has been recovered. It is almost always a doge, which is much more frequent than the Taningia, whose species has its highest concentration in waters of La Gomera.

The first sighting of a giant squid in Spanish waters of which there is documentation occurred in Canary Islands in 1861, with the famous encounter between a colossal Doge and the French ship Alecton. The encounter that led the Paris Academy of Sciences to examine the matter at their next scientific meeting, considering the possibility of recognising the new being in the scientific community.

One of the best captures took place in August 2011 when a film crew from the cetacean colony that lives in the waters of southern Tenerife located, floating dead about two miles off the coast, the remains of a large specimen. It was a female architeuthis who was sighted in front of the Los Gigantes cliff. The animal was in a good state of preservation, maintaining its original pigmentation, and measured about eight meters in length.

In June of 2019 we learned the sensational news: a giant squid had been recorded for only the second time in history ... and in Canary waters! The biologist Alejandro Escánez obtained the images at a depth of 500 meters from the oceanographic ship Ángeles Alvariño of the Spanish Institute of Oceanography (IEO).

The recording, about seven seconds long, shows the tentacles of a huge cephalopod approaching a camera, surely attracted by its luminescence. He takes the camera, shakes it, and leaves. The video was recorded in October 2017. After analysing the images once on the ground, they concluded that it was a Doge Architeuthis. The conclusions were presented in June 2019 at the International Congress of Marine Sciences of the University of Vigo.

Asturias, fishing ground of the kraken

The Asturian coast is the other major Spanish focus of interest on these species. In its waters there have been a large number of catches, with specimens generally larger and in a better state of conservation than those found in the Canary Islands. Since 1952 we have heard of more than seventy documented encounters with large cephalopods in Asturias, a

figure that represents 15% of the world cases known in this period, although there are much earlier accounts that already speak of the presence of these invertebrates in the area.

To talk about giant squid in Asturias, it is necessary to refer to three big names that are part of the small history of Spanish cryptozoology. We refer to the Carrandi fishing ground, the Kraken Project expeditions and the conservation and dissemination work carried out by CEPESMA and the Luarca Giant Squid Museum.

The Carrandi fishing ground is a pit located about 26 miles northeast of Gijón, just over two and a half hours by boat and it is rich in fauna. Along with the Ross Sea, located to the south of New Zealand, it has the world's largest concentration of these giants.

The reasons for this abundance are diverse. On the one hand, the close proximity of Carrandi facilitates the movement to the area for research and observation of cephalopods. On the other hand, it is an exceptionally rich region of fisheries. In recent years the area has seen an increase in the size of the fleet that fishes in the fishing grounds, which has also caused accidental catches of Giant Squid to increase.

The first campaign to study the giant squid lasted five days and took place in October 2001. The second, which lasted two weeks, was carried out in September of the following year. On both occasions, bad luck beset the expeditionaries: storms with waves of up to eight meters high prevented them from achieving their objectives. The results? More than a hundred hours of filming with which a documentary was edited, sonar images of what appeared to be a squid colony located between 600 and 800 meters deep and the capture of a dead male six meters long.

Fortune, or rather misfortune, has contributed to the fact that the number of specimens under study has grown in number after the massive appearance of corpses in the autumn of 2001 and 2003. The dates coincide with various oil and natural gas surveys carried out in the area by the Spanish Navy ship Hespérides along with other ships from the Repsol oil company. To carry out their research, they used echo sounders and compressed air cannons that produced disorienting acoustic waves in the squid, leading them to the beach.

In addition to these captures, some institutions - such as CEPESMA or the Spanish Institute of Oceanography of Vigo - carry out to collect as many records and specimens as possible. The sum of these variables explains why the Asturian coast is considered a kind of paradise for the study of these animals.

The second big name in the research on Asturian giant squids is the Kraken Project. With this name two expeditions are known that took place in the Carrandi fishing ground with the aim of filming for the first time a living giant squid.

Those who have been doing exceptional work for many years are responsible for CEPESMA (Coordinator for the Study and Protection of Marine Species) and the Museum of the Giant Squid in Luarca. The coordinator, created in 1994, carries out a fundamental work regarding the study, conservation and dissemination of knowledge about giant squid and other marine animals, both in her local area of influence and internationally. The Giant Squid Museum - destroyed by a storm in 2014 and whose reopening is scheduled for 2020 - gathered, for its part, what can be considered the largest and best world collection of large cephalopods.

Squids in the Strait
It is unknown if the giant squid has sedentary habits or if, on the contrary, any of the known species migrates during part of its life or occasionally. Everything seems to indicate that being a solitary species, its chances of migration are less than if we were dealing with gregarious individuals.

Historically, seven specimens of squid have been found in the area that stretches from the coasts of Cádiz to those of Valencia. There are not many sightings for 800 kilometres of coastline but we must get used to working in these conditions when we refer to large cephalopods.

This spatial distribution seems to suggest that these animals enter from the Atlantic through the Strait, following something similar to a migratory current or –at least– carried by the ocean current that penetrates the Mediterranean.

The last of the sightings that occurred in the area occurred in the month of October 2012, specifically on the Algeciran beach of Punta Carnero. It was a dux of About 70 kilos in weight and almost four meters in length, which was later transferred to the beach of Cala Arena, where technicians from CEGMA (Andalusian Marine Environment Management Center) took care of the animal. In the subsequent autopsy, carried out a little over a month ago, members of the Luarca Squid Museum participated.

In any case, we do not know the final destination of this possible migration that we are examining closely. Recently a nest of architeuthis has been discovered off the island of Sicily. Maybe the Atlantic squid migrate to the Mediterranean until they meet up with their Italian colleagues? It is a possible hypothesis with which we continue working.

Galicia, 2010, first capture of the kraken
In 2010, the Minchos VI trawler, based in the port of Celeiro (Lugo), captured a young specimen of giant squid in the Hercules Trench, 36 miles northwest of Coruña. The animal, belonging to the genus architeuthis, measured 8 meters and weighed 70 kilos.

It was the first capture of a giant cephalopod in Galician waters, which was a complete surprise to scientific circles. In the first week of April of the following year, the same boat captured another specimen in the same place. It was also a doge, this

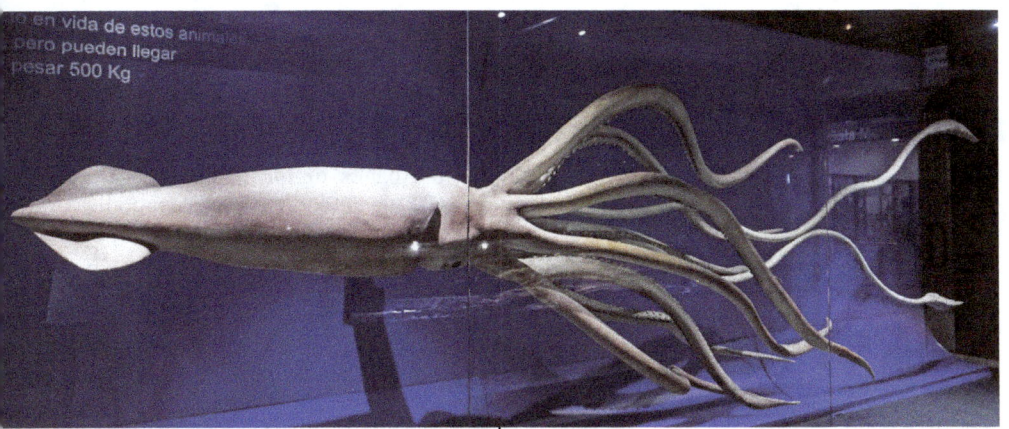

time with a weight of 48 kilos and a length of six meters, a female that lacked one of its tentacles and was caught in the vessels net at a depth of about 600 meters. From this date, in the Bares area, there have been four other encounters with giant squids (some weighing more than 100 kilos), almost all belonging to the architeuthis species, except for a *Taningia danae* found some 240 meters from depth that was transferred to the Spanish Institute of Oceanography of Vigo for study. It weighed 54 kilograms and its cloak reached 103 centimetres in length. Currently, the coasts of the Cantabrian Sea are those that concentrate the greatest number of cases of krakens appearing in Spanish waters ...

Much to do ...

Despite the data provided after the studies carried out on the most recently captured specimens, there is almost everything to do in the world of giant squid. These animals, which a little over a century ago were still considered the result of imagination, have not yet removed the halo of mystery and respect that accompanies them.

We know almost nothing about their adult physiology, their mode of reproduction, their hunting habits or their habitat. Recordings such as the one recently captured in Canary waters will help us to better understand these authentic colossi of the oceans.

While this is happening, from time to time we will continue to hear stories of terrible sea monsters barely glimpsed that will make our thoughts return unconsciously to the myth, the primal beast, the Evil One, the Kraken ...

ESCAPED AND REPORTED CROCODILES: CONTINENTAL EUROPE

Germany

A crocodile was caught in 1914 at Rastatt.

A small crocodile was caught in the Rhine, at Oberwesel, in 1929. It was brought to a zoo. (Bild 6 July 2001, p. 5)

1994, Neuss, Nordrhein-Westfalen, caiman Sammy escapes its owner while he takes it for a swim in a local bathing lake, police searches, bathing banned, weeks later, the tine frightened reptile is recovered. Made headlines the whole summer.

July 2000: Eislingen, Baden-Württemberg – alligator reported to police, 1 m long, was stuffed! (*Rheinpfalz* 25 July 2000)

2001: This was in the media world-wide, and I report mainly what the major German tabloid *Bild* wrote and one of the serious papers, the *Süddeutsche Zeitung*, and some regional papers. On 22? June, at a place variously given as Ketsch, Speyer and Altlußheim (but only the last is correct), cyclist Michael Berbner, 31, went for a pee at the banks of the Rhine when he saw a 5ft/1,5 m crocodile sitting motionless on a tree stump. He threw a lump of ground on it, it hissed and went into the water. Police considered this an authentic sighting and assumed the croc had been exposed by its former owner. *Bild* had pictures of a monster croc from the start, and the assumed reptile was soon a major attraction. The region was swamped by curiosity seekers, police patrolled with two boats, anglers warned the animal would eat all the fish in the river! (*Badische Neuste Nachrichten* 23 June 2001; *Bild* 23 June 2001, p. 6; *Bild* 25 June 2001, p. 3) However, neither police nor journalists nor anyone else saw anything, and the *Badische Neuste Nachrichten* 26 June 2001 reported that the police search was a failure. Yet it was generally assumed there was a croc in the Rhine, and mayor Hans Wirnshofer, 58, asked the authorities to ban public access to the river at Altlußheim. (*Bild* 26 June 2001, p. 3) The next sighting was on 26 June, this time at Eltville near Rüdesheim – 113 km down the river. Two observations were reported early that morning, both in the vicinity of Mariannenaue island. Capitain Karl-Heinz Weinert, 58, of the ship "Sinn" at 6.24 hours spotted something in the water: "I at first

shutterstock.com · 35254030

Ulrich Magin

thought it was a log. Yet that log swam toward the island and went ashore. Then I recognized the crocodile. It must have been 10ft/3m long!" The second observation was not detailed in the press. *Bild*-expert vet Dr Florian Brandes explained that the animal must have drifted in the river for 113 km. Police used helicopters, boats and 20 persons. Gunther Köhler, head of the Reptile Department of Frankfurt's famous Senckenbergmuseum was full of doubt, however: "Crocodiles move a maximum of 30 kilometers in 2 to 3 months." An animal park in Hesse was ready to house croco, should it be caught. (*Badische Neuste Nachrichten, Badisches Tageblatt* 27 June 2001; *Süddeutsche Zeitung* 28 June 2001, p. 12; *Bild* 27 June 2001, p. 3) After these three sightings, a previous observation was now reported in the *Allgemeine Zeitung* of Mainz. They had received a woman's call who said she had observed the crocodile already on 9 June near Heidesheim. Police

did not know of the sighting, as spokesman Lothar Neumann explained. The woman had spotted the crocodile only 5 m from the shore when she took a walk with her cousin. The press began talking about "a second crocodile". (*Rheinische Post* 28 June 2001; *Badische Neuste Nachrichten* 29 June 2001) The usual fuss followed: police and tourists combed the region between Mainz and Bingen in helicopters, boats and cars. Camera crews and journalists crowded the place. (*Badische Neuste Nachrichten* 28 June 2001; *Bild* 28 June 2001, p. 1 & 3), finally even divers – all failed to detect any sign of the creature. (*Bild* 29 June 2001, p. 3) Bathing and swimming was banned by local law. A police spokesmen explained the danger were not yet over! They continued to look out for a 1.5 m crocodile. As no new sightings were reported, *Bild* – again depicting a monstrous reptile – brought tips on how to behave once you encounter the monster (*Bild* 30 June 2001, p. 7) and interviews with the various hobby crocodile hunters. (*Bild* 2 July 2001, p. 6) There were no sightings until 2 July, when at 22.30 hours captain Hartmut Jeschke, 57, and 2 members of his crew of the tanker "Adria" spotted the crocodile in a dark but clear night near Bingen. They first heard knocking sounds, then saw the animal drift past the ship's hull and then swim toward the island Aue. "I saw the gigantic crocodile. About 2 m long. It had thick, brown scales", the captain said. And helmsman Hein Schneider, 65: "It swam past the ship. It whipped the water with its tail, vanished in the eddy of the ship's screw in the direction of the Rüdesheimer Aue. It was really frightening." Police spokesman Helmut Oberle thought the account was credible. Again, police searched with helicopters and boats - without result. (*Süddeutsche Zeitung* 4 July 2001, p. 12; *Bild* 4 July 2001, p. 8; *SDR 3* news 3 July 2001; *Badische Neuste Nachrichten* 4 July 2001) On 4 July, water policemen saw the crocodile near Bad Geisenheim in the Rheingau. It swam lazily in the water near the bank, but, when caught by the police, proved to be only a carved wooden crocodile, 70 kg, 2 m long, complete with head, scales and fierce eyes. Police spokesman Norbert Hübsch: "Someone is poking fun at us, and must have worked real hard for it." Student Robert Koch, 27, is still on hunt with a canoe and sausages as bait. (*Bild* 4 July 2001, p. 3) Only a few days later, some people taking a walk found a plastic crocodile – no place mentioned.

October 2002, Berlin: police is informed of 1.3 m caiman in high-rise building, they catch it, but it is a pet. *Bild* 11 October 2002

August 2002, Edingen, Baden-Württemberg: police is informed a croc swims in river Neckar, 80 cm long, police recovers the croc. It is made of plastic (*Bild* 31 August 2002)

May 2005, gravel pit near Biberach, Baden-Württemberg: 2 anglers report 1.5 to 1.8 m (5 to 6 ft) crocodile in lake, police concludes it is only a giant catfish (*Die Welt* 31 May 2005)

July 2006, Nürtingen, Baden-Württemberg: 50 cm Yacare caiman escapes from cage of owner Oliver S., 39, , goes into river Neckar, police searches for a week, neighbour finds it on his green, owner goes and simply takes it back. *Bild* 11 July 2006

October 2007, Schwarzenfeld, Bavaria: in the Grüner See, a bathing lake, several people, including anglers, report 2 m crocodile, police and fire fighters search in vain. Police suspects that a beaver has been seen. *FAZ* 12 October 2007 + all major papers

Early in May **2008**, several youth observed a crocodile crossing Lake Schwielow, near Berlin. Later, additional reports came from the river Havel south of Werder. Jörg Lippert of the Federal Land Environment Department was sure the sightings could be explained by beavers: "If you do not know these animals, it is easy to mistake them for a crocodile." He explained that a beaver had taken residence just where the sightings had been made. (*Tagesspiegel* 18 May 2008).

An 80 cm crocodile was spotted in an arm of the Innerste River at Hildesheim by employees of the city's water department on 26 May. After a long search, Police explained on 28 May **2008** that they had not found anything. No further search was planned (*ddp*, 28 May 2008). On 3 June, a snapping turtle, 50 cm long and weighing 15 kg, was caught at Henneckenrode – and the press reported that the mystery was now explained (*Focus online* 3 June 2008).

On 29 June 2008, at 1700, children playing at the Friedheimer See near Cuxhaven spotted a crocodile lying still in the water. It then jumped out of the lake to catch a bird, claw marks were discovered in the sand on the bank. Police investigated but found neither the croc nor the alleged traces of it (*www.presseportal.de/polizeipresse/ pm/68437/1220809* 2 July 2008).

On 24 August 2008, at 20:39, a woman called police after she saw a "giant lizard" crawl along the Saarwellinger Lake, Germany. A few days later, police located a man who owns a 1.4 m iguana which he takes for a swim in the lake (*Saarbrücker Zeitung* 27 August 2008).

A witness observed a dead crocodile in the Nagold (a minor German river) at Pforzheim on 13 July **2009** at 2100. He said it was 60 cm/2 ft long and was drifting at the surface. The man stood at the central Ludwigplatz, the creature was about 200 m from him. When police arrived, they found no crocodile and assumed it had been washed away by the river (*Stuttgart Journal* 16 July 2009; also in *Pforzheimer Zeitung*).

In a pond near Pressath in the Upper Palatinate, Bavaria, two 15-year-old girls spotted a crocodile on 14 August **2009** at 9 pm after they heard a noise in a bush. When they neared, the 1 m / 3 ft reptile went into the water and disappeared. When it was suggested the girls had seen a beaver, they denied: their creature had had a pointed snout, brownish green colour and two humps on the head. As the police thought the witnesses were credible, the lake was declared off limits, police went there with 55 men, a police woman controlled the bank with a police dog, water safety searched the surface using 5 boats, trying to stir up the creature with their paddles. As all this lead to nothing, the lake was declared safe the same afternoon, and bathing was again permitted. Yet police will continue to keep an eye on the water. (*Kölner Stadt-Anzeiger*, 16 August 2009)

A tourist snapped a photo of a crocodile in Lake Malchin, near Berlin, although the photo is too blurry to be of much use (*Nordkurier* 16 September **2009**).

2010 Five kids from Bochum reported two 1.80 m crocs. One was basking on an island in the river Ruhr, the second approached swimmers in the water. *rp-online*, 9 July 2010

In the Mittlerer Klausensee at Schwandorf, a pond rather than a lake, a 44-year-old woman swam from the banks to her inflatable mattress early in July **2012** when suddenly a 1 m long crocodile with a long tail swam directly over her. She had a 7 cm / 3 in wound as evidence. Then, on July 7, a man walking along the bank alarmed police when he saw the reptile basking on the shore. Many police searched the lake and the environment after that, but failed to arrest the crocodile. Nuremberg Zoo said such a creature posted no more danger than a dachshund, though. Police spokesman Thomas Hecht thought the crocodile was no phantom, but had been released by an exotic pet owner, as in the region, about 5 years ago, there had already been crocodile sightings, this time in the Grüner See of Schwarzenfeld. At that occasion, no reptile was ever discovered. Sceptics assumed the hiker had only seen a beaver. (*Stern.de* 9 July 2012; *web.de* 11 July 2012) After a week of headlines, police explained on July 20 that the crocodile had been identified by wildlife cameras as a beaver (H*amburger Abendblatt, Welt Online, Frankfurter Rundschau, FOCUS Online, Spiegel Online*, and 312 further media, on 20 July 2012) In direct contrast, a second police statement published in the *Stuttgarter Zeitung* (26 July 2012, p. 9) stated that the "crocodile" had "finally" been positively identified as a 45 cm long, harmless bearded dragon. On 26 July **2012**, the *Rheinpfalz* (credit: Alfons Magin) reported that a 67-year-old motorist had caught a 34 cm bearded dragon on the pond's banks, leading police to claim this had caused the sightings (but bearded dragons can't swim …). The crocodile had first been seen on July 7.

In mid-August 2012, seven unsuccessful searches by police hunted for a crocodile that had been spotted by an angler in the Großkaynaer See, near Großkayna, Saxony-Anhalt, an artificial lake that fills a gap in the ground left from coal extraction. (*BILD* 24 August 2012)

On 4 September 2013, a woman encountered a 50 cm crocodile in a stream running into the Elsenz river close to Eppingen, Baden-Württemberg, Germany. Police investigated and found a plastic toy (*SWR Landesschau* 5 September 2013; via Markus Hemmler)

Around the banks of the Luegsteinsee at Rosenheim in Upper Bavaria, signs were erected by persons unknown warning that a dangerous crocodile had been seen in the lake. It was assumed that no such sighting had been made but that someone unknown attempted to keep tourists away from the lake. However, the village next, Oberaudorf, used the signs to promote a "search for the escaped baby alligator" and to lure more tourists to this beauty spot. (*OVB online* 9 August 2014)

Austria
At Vienna, fire fighter divers fished a 70 cm crocodile from the Danube Canal after passer-bys had observed it and brought it to Schönbrunn Zoo. (*20 Minuten*, Berne, 20 July **2001**, credit: Andreas Trottmann; *Bild* 20 July 2001, p. 12, here the animal is called a caiman)

May 2006, Silbersee near Villach – hikers report a 1.5 m / 5 ft caiman in the lake, long search, no results (*Neue Kärtner Tageszeitung* 12 May 2006, *Kleine Zeitung* 11 May 2006)

May 2006, river Leitha, Burgenland: several hikers report large crocodile jumping from bank into river, police searches, in vain. ORF 3 May 2006

At the end of August 2012, two 11-years-olds were taking a bath in the river Drau near Sachsenburg when they were attacked by a crocodile on a sand bank. The creature allegedly snapped at their shoes. The shoes were sent to the Natural History Museum in Vienna to search for DNA, but none was found. Then, at the beginning of September, two German tourists (or a single male German tourist, depending on the source) on a cycle tour through Austria reported a sighting of a 1.5 m/ 5 ft monitor lizard or iguana at the very same spot. It should be pointed out, however, that Austrian newspapers were full of reports on the fruitless search for "crocodile Sachsi" in the river. In mid-September, the museum reported that no DNA had been found on the shoes. A spokesperson said it was difficult to answer the question if there had been DNA on the shoes which had now disappeared, as to many people had handled the evidence and possibly obliterated all traces. (*ORF* and *oe24.at*, 4 September 2012; *Kleine Zeitung* 17 September **2012**)

France
I found what might be one of the first European "escaped crocodile" episodes recently. On 17 July **1858**, the *Chicago Tribune* reported (this is a pay-per-view-paper, so I retrieved only fragments: "A Rival of the Sea Serpent. The Silver Lake sea serpent has a rival in France, which the Paris papers are discussing. A correspondent of the *Patrie* gives the following of him: … Seine has also written a letter upon the subject, in which he says: Ten persons, at least. At Vitry, the presence of the monster." - a ' e (the cruel crocodile,) has devoured a dog in the presence of his (the dog …" This is also alluded to in the *New York Times*, 26 July 1858, p.2: "But perhaps you have heard

of the crocodile of the *Patrie*. The other day that profoundly serious journal asserted that a crocodile had been seen in the Seine near one of the Paris bridges, and you may judge of the surprise of the land-lubbers who make up the population of Paris? The crocodile of the *Patrie* is one of the lions of the moment." It was also referred to in the 11 April 1860 edition of the New York Times (p.9), in an article on gossip and propaganda in France: "The journal which commenced at Paris the agitation (…) – the *Patrie* – [is] none other than the journal which discovered a crocodile in the Seine …" From that, I reckon that there had been several reports about a crocodile in the Seine, at Vitry, a suburb of Paris, which were mainly reported by a single newspaper.

Italy

At Mantua, in the church Santa Maria Vergine delle Grazie (1399–1406) in Curtatone, a crocodile mummy is on display. Allegedly, it was caught around 1500 in the reeds of the river Mincio (Cordier 1986, S. 91–93; Centini 2001, S. 26; Neuteboom 2003; Schenda 1995, S. 182).

Carlo Amoretti reports, in 1814, a stuffed crocodile on display in the church of St Mary on the Sacro Monte near Varese which is said to have been caught in a valley near Lago Maggiore between Breno and Lugano (Anon. 1994q, S. 37–39; Centini 2001, S. 26 –27; Castiglioni, 2002, S. 6).

In 1865 it was reported that a croc mummy was also displayed in the church Santa Marta, in Como, which was later housed in the city's library – no trace of it remains today. (Castiglioni 2002, S. 6).

In July 1999, a search for a 6ft crocodile in the delta of the Po River, Italy (*Fortean Times* 127, S. 10; *Fortean Times* 251, S. 80)

Early 2000s –long crocodile hunts in Lago di Massaciuccoli, Tuscany, can find clips if

Crocodilus   Crocodyll

required

The *Corriere della Sera*, on 17 October **2000,** reports a "croc alarm" art Monza. An advocate saw the 50 cm croc in the park in front of the courts, and he immediately informed police who found the reptile … made of plastic.

Early June 2000 saw a hunt for caimans in lago Maggiore. Police had raided a house in Fondotoce and found an caiman which was housed in cruel circumstances. It was claimed that the animal had been kept with 7 more, of which no trace remained. This is how the rumour started they had been released into the lake. (Anon. 2000a; Crenna 2000; Offreddu 2000). No sightings were reported.

A small alligator was observed early in August **2002** in the Valle Intelvi near Lake Como. Two anglers saw the 50 cm alligator at the confluence of the rivers Brentana and Telo. Angelo Bernasconi said: "I had enough time to observed the small reptile. It was no longer than half a metre, it had a short, trapezoid head armed with teeth. It noticed us and dived and vanished at the bottom of the river Tela." Foresters searched but in vain. Aita, F. (2002): Un alligatore ‚smarrito' in Valle Intelvi. *Il Corriere di Como*, 11 August 2002, S. 4; Aita, F. (2002): Un ‚task force' per l'alligatore. *Il Corriere di Como*, 14 August 2002, S. 4

Convento di Santa Fiora, Tuscany, Italy: This church preserves the remains of a lake monster, which recent analysis found out was the cranium of a Nile crocodile (*La Stampa*, 16 December **2002,** p. 3)

The tentacle of a 30m/100ft squid was found in lake Vico in Central Italy in August **2004,** and brought to a forensic laboratory, where it disappeared. Then Sergio Stivaletti came forward to say it was part of a prop he had used to film a monster movie there. Police of

Viterbo were relieved, they had had to deal with UFO and **crocodile** sightings that summer and were happy there was no real monster in the lake (*Repubblica* 20 August 2004, p. 1 and 7).

On April 26, 2010, a passer-by spotted a crocodile in lake Falciano del Massico, Caserta, Campania, which is part of a nature reserve, and the lake has been closed off. The incident made headlines in all major Italian newspapers, often as "Loch Ness in Italy." The witness, who used to work in a circus where he looked after the reptiles, is certain of what he saw. Early in May, remains of a two-kilo fish which had been torn to pieces were found in the lake and footprints near the spot where the creature was seen are being investigated. As usual, it was assumed the croc had escaped from a circus or had been exposed there by a private owner. Also, many criminals in Italy keep crocodiles for prestige (one had been "arrested" only a few months ago in Naples). (*Provincia di Como* 3 May 2010, p. 7; *Corriere del Mezzogiorno* 27 April 2010)

In a river near Foligno, Umbria, Italy, a crocodile was seen in March **2011**. (*tuttoggi.info* 8 March 2011)

A large crocodile was repeatedly observed in the Oasi del Nervia, the mouth of the river Nervia near Ventimiglia, Italy, in February **2012**. (*Il Secolo XIX* 8 February 2012; *Riviera24.it*, same date; *Sanremonews* 9 February 2012)

A sighting of a crocodile in a lake near Caserta, Italy: *Solo News* 12 March **2012** – **see 2010**

At Turano Lodigiano, situated close to the river Adda between the Po and Lake Como, a crocodile was hunted early in May **2013**. A witness had reported, on afternoon of 9 May, that he had seen the large reptile emerge from the Muzza canal that bisects Turano.

KROKODILE.

1. Hechtkaiman (Alligator lucius). Länge ca. 5 m.

2. Gangesgavial (Gavialis gangeticus). Länge ca. 6 m.

3. Nilkrokodil (Crocodilus vulgaris). Länge 7—10 m.

Brockhaus' Konversations-Lexikon. 14. Aufl.

The Forest Guard immediately organized patrols along the canal banks, yet most of the villagers remained quiet and sceptical. There was one additional report about a reptile, although not of great size, that had eaten a moor hen. The Forest Guard couldn't find a trace of the croc. They intended to search further in the following day, and a school trip along the Muzza was cancelled. The creature then faded from the news. On an earlier occasion, in July 2002, near Castelnuovo Bocca d'Adda, eyewitnesses had seen a large panther in fields of corn. A hunt was organized, but failed to find a trace of it. Regarding the crocodile sighting, the newspaper asked: "Is this a big hoax or really an alligator?" (*Il Giorno* 10 Mai 2013)

Lake Orta has ancient legends as well as modern rumours (Alessi, the designer who lives there, told reporters there is a monster on the lake – everybody knows it!). From July till August 2014, rumours circulated at Lake Orta that there were crocodiles in the lake, one, two, even three. Locals remarked how there were fewer ducks than usual, and everybody seemed to know that someone else had seen a croc of unspecified size in an unknown area of the lake. The testimonies, remarked *La Stampa* (4 August 2014) "as in any urban legend differ in detail and are fun. The crocodile has a variable size depending on the location of the sighting. It is large near Omegna, a little smaller in Pella and only slightly larger than a lizard in Gozzano." It was claimed a helicopter was searching for the beast, but this was the famous photographer George Gnemmi, specializing in aerial images, and he did not even knew of the croc. Moreno Lubelli, of the public ferry lines of the lake, said he had heard so much about the croc he almost believed in it, however: "the only thing we noticed were the coypu, which are also of a certain size." Mario Savoini of Sub Novara Laghi explained that even if someone had thrown a croc into the lake, it would not survive winter temperatures.

Spain
In August 1983, a 50 cm caiman was captured in the Rio Guadalhorce (Fernández, 2003)

A crocodile was hunted in the laguna El Portil, Huelva, in July **1991**. (Efe 1991; E.P. 1991

In November **1998**, policemen equipped with sonar searched for a crocodile that had been observed in the Lago de Campo de Naciones, at the Sevilla Expo. (Anon. 1998)

Also in November 1998, police searched for a crocodile that had been seen, by gardeners, in an artificial pond in the Parque Juan Carlos I. in the centre of the capital Madrid. (Cándido 1998) (taz 8. December **1997**, p. 20)

Also in Madrid province there was a long series of sightings of crocodiles in the reservoir of Valmayor in the summer of **2003**. These report even managed to get into the pages of the *Fortean Times*. On 2 June 2003, two locals observed one or two 3 m / 10 ft. long crocodiles in the water of this 12 km long reservoir which supplies drinking water to Madrid. One of the witnesses was a woman walking her dog. Policemen and Guardia Civil patrolled the lake with helicopters and boats. Naturalist Luis Miguel Domínguez visited the site of the encounter and explained the ground traces left by the animal as traces of wild boar. (Serrano 2003a)

On 6 June several witnesses, including a policeman, observed two crocodiles in the lake. (Serrano 2003b) Ten days later, authorities were still searching night and day for the creature, but without result. Domínguez suggested that the sightings in the water were actually observations of nutrias. (Anon. 2003; Lillo 2003; Ussía 2003)

The crocodiles were the summer topic in Spain in this year, news reports also appeared to report that nothing new had actually happened – even this was news! (Serrano 2003c) And the Valmayor crocodiles were called "the Nessies of Spain". (Meseguer 2003).

One year later, it was still news that no new observations had been reported. (Aguilera (2004; Anon. 2004)

In April, July and September **2007**, on four occasions, crocodiles were caught or observed in the vicinity of Jerez, though each time near a reptile farm. On 9 April a man captured a crocodile near a stream at Puerto Real, in July, a 1.3 m / 4 ft. crocodile which had escaped from the farm was caught, in September a crocodile more than 1 m in length was spotted in the Laguna de Torrox. A hunt lasting over a week failed to capture it. (Camacho 2007 a, b)

In February **2013**, police were searching for a crocodile in the waterways near Mijas, at the Costa del Sol, Spain. (*Fortean Times* 301, p. 8)

Greece
Reservoir near Rethymnon, Crete. On 6 July **2014**, a team of local fire officers on patrol spotted a two-metre long crocodile in the lake. (According to a German press report, there were further sightings on the same day.) Two reptile experts from the Heraklion Natural History Museum were expected to help police capture the animal. Police were also planning fence the dam. Athens Press Agency Ana said authorities were concerned by rumours of the presence of a second crocodile. "Clearly crocodiles do not occur naturally here, so the owner probably wanted to get rid of it," a local police official told news agency AFP. There were considerations whether the crocodile, when caught, should be brought to a local zoo or whether it should remain in the lake, as it appeared acclimatized, to be used as a further tourist attraction! (*skynews.com.au* 8 July 2014; *Rhein-Sieg-Rundschau* 11 July 2014, p. 27) The artificial lake is the Amari Dam Reservoir. According to the website http://www.cretanbeaches.com/Lakes/Lakes/potami-dam-lake-amari/: "The lake of Potami is shaped by the Potami dam, in Amari plain. It was built in 2008. The dam is located in the verdant valley of Amari, 25km south of Rethymno. The quite new lake has a capacity of 23 million cubic meters and is expected to become one of the most important wetlands in the southeast Mediterranean. Already, many species of birds and animals have appeared in the area. Visitors of the wider area visit the dam and admire the beautiful scenery. On July 2014, the dam came to light when two fire fighters accidently saw a crocodile, 2m long, on the shore! It was apparently freed by someone unconscious. Various Services participated in the capture of the lovely reptile."
Crocodile sightings in a reservoir at Rethymnon, Crete. (*Die Rheinpfalz* 11 July 2014, credit: Alfons Magin)

Cyprus
The Kouris Dam reservoir, completed in September 1988, is one of the largest in Cyprus, and is fed by several rivers. Located 15 km north of Limassol, it was the site of a minor excitement in **2005** when somebody reported that (a) small crocodile(s) illegally imported from Egypt had been thrown into the lake. Since then, there have been many sightings of „a strange creature", and local newspapers have reported countless observations of what was dubbed the „Cyprus Loch Ness". „I watched this serpent with my own eyes, this was no mistake", a witness is quoted. Recently, it appears there were several complete searches of the lake this year (one source says eight), but no sign of an unusual creature was found (*Famagusta Gazette* 28 October **2008**; *New York Post* 27 October 2008; *Reuters* 24 October 2008).

Poland
In March, a 1.5m/5ft, 20 kg crocodile frozen to death was found in Poland. (credit: Andreas Trottmann 4 March 2011 **SORRY; I LOST THIS CLIPPING!**)

UK
Swansea, Wales: On 27 April **2008**, fisherman Steve Jenkins saw a crocodile in Morfa Enterprise Zone's Pluck Lake, a 20ft deep pool in the middle of a city business park at Swansea. Jenkins was walking his dog around the lake, when he saw what he first thought was a log in the water but: "It was definitely a crocodile. There was a white van submerged in the water and it swam over the top of it so I had a good look. It was a metre long and had a long tail. I've been ribbed mercilessly since I reported it to the police – people whistle the tune of Crocodile Shoes when they see me. But I don't care what people think, I know what I saw." (*Western Mail* 30 April 2008; *Mail on Sunday* 30 April 2008) There was already some talk of promoting the crocodile as a Welsh Loch Ness Monster.

Sightings of crocodiles in the Thames, England – *Fortean Times* 300, p. 8

Crocodile sightings in River Avon, at Bristol, UK – *Fortean Times* 313, p. 10

Bristol crocodile, new sightings: *Fortean Times* 318, p. 20

CEYLON CROCODILE *var. from Seba.*

THE YETI OF THE HIMALAYAS

by
Saarthak Halder

The yeti, a huge, gigantic, shaggy, hairy, bipedal, ape-like beast which roams the range of the Himalayan mountains in Asia has been seen by many Himalayan residents (mostly by citizens of India, Nepal and China) and although it is believed by some to be a mythological deity .

Ancient people and the old settlers from Tibetan culture who resided in the Himalayas worshipped these glacial beings and believed in them as rulers of the mountains and forests, and there are many stories told about them, in this guise.. They also believed that they have mystical healing ability. It is said that when anyone damages or tries to ruin the snowy peaks, they become angry but other times they help mountaineers to find their way in blizzards or avalanches.

In another culture of ancient Tibetans, people organized spiritual ceremonies to honour their yeti deities, and these were carried out in order that the people would receive the protection of these glacial beings. These ceremonies involved the sacrifice of an animal.

But las time went by perceptions changed, and people started seeing yetis as devilish, demonic figures.

In Nepal, they call these demonically perceived yetis "Rakshasha" (demon in the Hindi language) and believe that they are violent creatures who attack livestock and humans and their legends depict yetis as a dangerous creature and one to be avoided if you don't want to welcome misfortune in your life.

ADG PI - INDIAN ARMY
@adgpi

For the first time, an #IndianArmy Moutaineering Expedition Team has sited Mysterious Footprints of mythical beast 'Yeti' measuring 32x15 inches close to Makalu Base Camp on 09 April 2019. This elusive snowman has only been sighted at Makalu-Barun National Park in the past.

10:43 PM - 29 Apr 2019

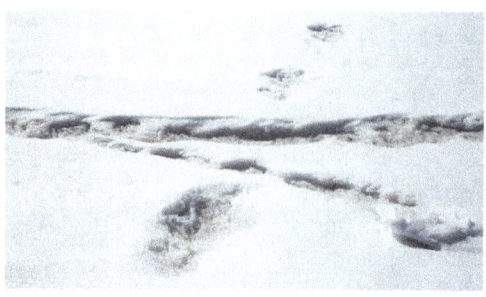

It's still a mystery whether yetis are actually dangerous or not.

The Indian army has posted some pictures

of footprints on twitter on 29 April 2019, claiming them to be those of the Yeti. It's claimed that the discovery has been made on 9 April. It attracted huge number of people worldwide. I was a bit surprised when I saw that.

They found the prints in Makalu base camp in the Himalayas and it's been said that elusive snowman had been sighted at Makalu - Barun national park in the past. The Makalu - Barun national park is in Himalayas, not so far from Nepal.

This claim was mocked by many but the subject experts were bit hopeful and I think perhaps this can rekindle hope in the existence of the Abominable snowman.

Experts and theorists suggested that these glacial beings can be found anywhere on the Himalayan mountains and can reside in every terrain.

But these mysterious creatures have never been found properly and only we only have anecdotal evidence of sightings of and photographs of some mysterious footprints as fragile evidence.

Why they were never been found properly?

Why they are so hard to find?

There are many reasons which have been hypothesized .

The main reason could well be that perhaps, there are only a very limited number of these creatures still in existence and that they are so widely distributed over the whole Himalayan mountain range that their population density is extremely low and so hence it's humanly next to impossible to explore and find them.

Something else which makes their discovery more hard and toilsome is presence of caves and forests.

Subterranean caves are interesting feature of Himalayas, and more interesting is the presence of many different types of caves.

There are also man-made caves like mustang caves or sky caves of Nepal; a collection of some 10,000 man-made caves dug into the sides of valleys in the Mustang District of Nepal (below), but it's impossible that yetis will make these their dwelling place. But there are many other naturally made limestone caves hidden in mountains, ice caves hidden in glaciers and hole-like caves in the sandy cliffs of Himalayan mountains. There are temporary caves too, like the fascinating ice caves formed within the glacial bodies. These are formed in those parts where

temperature goes subzero in winters and temperature rises during summers.

There are many different type of caves which are still fully not discovered and waiting to be explored. Those caves could well be great hiding place where no sensing equipment can reach.

Forests can also play a major role in providing hiding places for many kinds of creature including giant glacial beings.

The forest vegetations and ecosystems can be broadly differentiated into different types such as:

- Tropical,
- subtropical,
- temperate and
- alpine.

These form various biotopes such as:

- Himalayan subtropical pine forest,
- Western Himalayan temperate broadleaf forest,
- Western Himalayan temperate sub alpine conifer forest
- and Himalayan subtropical broadleaf forest

As the name suggests, temperate broadleaf and sub alpine forestation prevails in temperate regions and subtropical broadleaf and pine prevails in subtropical regions. The temperate broadleaf covers 55,900 km² area, temperate sub alpine covers 39,700 km² area, subtropical broadleaf covers 38,200 km² area and subtropical pine covers 76,200 km² area.

These are only examples as there are many different types of forest ecosystem also

and I think the above given data is enough to make you realize that what a tremendous amount of land is covered in forest which is extremely hard to explore whole.

There are many protected areas which are conserved as sanctuaries or reserve parks but most of these areas are covered by forest which are not or very barely inspected and surveyed, hence there is a strong possibility that these ecosystems can serve as a comfortable hiding place to beasts like the abominable snowman.

The abominable snowman itself also doesn't like to come out and meet face to face with human beings. In all the stories, religious writings and mythologies they are depicted as very intelligent beings who can understand a wide range of circumstances and human behaviour as well. So perhaps they consider humans as threat, possibly because they may have witnessed humans destroying habitats and hunting and killing other creatures.

There is also a big possibility that it's their behaviour which is incorporated in their genes to hide and lead a reclusive life. Many different creatures have evolved this type of behaviour so that they can flourish easily without staying in constant fear of outside predators and can also hunt easily. Perhaps yetis subconsciously follow this congenital behaviour which is carved in their brains by the evolution of their genetics.

Now, as we know, the yetis are supposedly beasts with fur, which means that they may camouflage themselves and can conceal temporarily themselves from humans. It can also be done by burrowing in the snow and hiding there transitorily and can mask their location, identity, movement and tracks. By doing this they can avoid predators and pounce on their unsuspecting prey.. But as their feeding habits are a mystery and we know that most probably they will be omnivores, so we don't know whether they can attack humans or just feed on very small animals.

The tales and lore confuse us more than giving an straightforward answer. Some ancient tales say that they are very nice and kind towards humans and only attack if humans try to damage their habitat or try to threaten them. There is another group of people who say that yetis are violent, tormenting beings who enter their village and try to kill and eat the livestock (I think perhaps these accounts get confused between yetis and brown bears, but who knows?)

Even though, it's tremendously hard to find them, there are numerous cases of sightings where people have seen yetis with their naked eye and there are proofs too like bones, hairs and majorly, the footprints which has been claimed as those of yeti's.

The history of the Yeti began with the Himalayan folklore, telling tales of a very large, muscular, man-like creature covered in fur, living in the snow-covered mountains of Asia. People at that time called it various names but it became popular after 1800s.

B.H Hodgson, a British naturalist and ethnologist working in India and Nepal reported that his local guide spotted a giant hominid-like creature covered in thick fur.

In another case, group of five miners decided to stay in their homemade cabin on a summer night. When they tried to get some sleep, a group of crazed hairy apemen started screaming and launching heavy stones at their cabin. They survived somehow and one miner reported he saw an apeman standing in a distance . He even wrote a book on that incident capturing his every experience (Perhaps the apemen were infuriated on some of their actions).

Laurence Waddell, a Scottish explorer, and army surgeon in India and collector in Tibet also reported his guide's description of a large apelike creature that left footprints.

Sightings increased during 20th century when many people started climbing the mountains.

N.A Tombozi, a Greek photographer and geologist also sighted a bipedal creature making big footprints on the way but he was sceptical about yetis hence he thought it was some hermit with big body.

A major and renowned evidence was the photos of yeti's big footprints taken by a very respected mountaineer, Eric Shipton, in the Himalayas. Some said that its the best evidence for so far for the yeti, while others argued it had been made by some other creature and it got distorted by

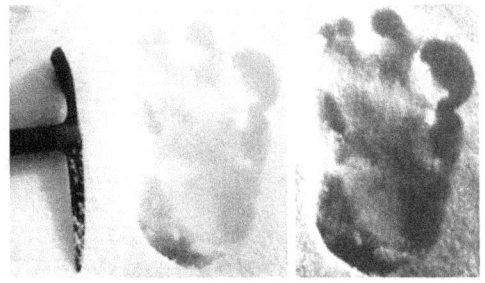

melting, but after seeing and examining that image, experts and I also don't think it has been distorted by melting.

Sir Edmund Hillary and Tenzing Norgay also, reported seeing large footprints while scaling Mount Everest. Tenzing Norgay was sceptical about yetis but he said he has seen the big footprints and he also admitted that his father said once that he has seen one.

In the Garo Hills also, there have been reports of yeti sightings.
]
A TV show host, Josh Gates, also claimed that he has found mysterious footprints in snow near a stream in Himalayas.

The sightings are not limited. There are numerous sightings which have been told by many mountaineers, explorers and residents of the villages near the Himalayan ranges.

There are many supposed pieces of evidences which have been presented, like hairs and bones. Most of the evidences were proved to have false like the Pangboche hand (a replica of which is pictured below) from a Buddhist monastery in Nepal. Eventual DNA analysis showed that it was a human hand and most probably it was that of a monk.

Another evidence was a scalp which was supposed to be yeti's and was borrowed from monastery by Sir Edmund Hillary and was tested and was later found to be the scalp of some species of serow.

In another case, some Nepalese gave evidences of yetis as skulls which was later confirmed as Asiatic black bear's skulls.

On 25 July 2008, BBC reported that hairs collected in remote the Garo Hills has been analysed and has been found that the hairs came from a Himalayan goral (below).

In 2013, geneticist Bryan Sykes asked the institutions of the world to send every yeti evidences they had and received 57 samples like hairs, teeth, bones and other tissues. Most of the samples turned out to be from cows, horses and bears.

In 2017 another team of scientists analysed nine other specimens which were found in monasteries and caves and the research revealed that eight were from brown bears and the ninth was from a dog.

But these discoveries are not enough to disprove the existence of yetis, as there are still many unexplained evidences of yetis which are a mystery and perhaps those mysteries can pave the way toward the existence of yetis.

Another evidence was found during a mission funded by Thomas Slick Jr, an inventor, businessman and adventurer, to investigate the yeti and on the expedition they found faecal matter of a creature whom they were calling a yeti and scientists proposed that parasites in the faecal matter may provide some clues about the creature but faecal analysis found a parasite which was apparently unknown and unidentifiable just like the creature.

Another evidence was found by Angelo Jackson, an educationalist and explorer, who photographed many footprints in Himalayas and among those footprints, there were large footprints too which could not be identified.

The yetis are very hard to find and evidence also lacks the power to bring the proof of the yeti's existence in front of the whole world. As we now believe, yetis are very intelligent beings, so perhaps it hides its own evidences in nature, so that they don't get caught. Or there are people out there who try to conceal the existence of yetis and save them from the outside world by manipulating the evidence.

There is another major point to think about and it is that many governments also tried to hunt yeti and perhaps they found evidences which they never revealed.

US and Nepalese government worked together to regulate the hunt of yeti and experts of both the countries searched the Himalayas thoroughly, there were some rules too in which one rule was "Don't tell the press" until the government agrees and gives the permission.

Russian government also took an interest in yeti in 2011-2012 and organized a conference and a hunt with Bigfoot experts and biologists. A renowned biologist named John Bindernagel claimed that he saw evidence of the yeti and said the yeti not only exists but also dwells in trees and caves and makes shelters.

There are many theories by experts that try to explain:

- what yetis really are?
- how they reside in those conditions?
- and how they evolved?

There are also numerous theories which are hypothesized by experts and by me.

Around 300,000 years ago, a genus of ape called Gigantopithecus coexisted with *Homo sapiens* near the southern part of China. Their species was *Blacki*. *Gigantopithecus Blacki* was a herbivore which relied on plants like cereals, Bamboo, cotton, Beets, chlorella etc. Figs were also an important dietary component. They weighed 300 - 400 Kg and were approximately 4 meters in height.

Somehow they were slightly suited for cold climates as inhabiting in cold climates require more body mass, thick fur to keep warm, small ears to inhibit shedding of body heat.

They were apparently good in adapting to different environments as the youngest fossil evidence show many differences from early members of the same species, depicting that they adapted themselves to better suit the changing habitat and climate.

There has been speculation that they had gone extinct because of the retreat of the forest and that this occurred due to increasing seasonality and monsoon strength and early human activities. As we know now *Gigantopithecus blacki* were good at adapting to these changes so that they may have adapted themselves and survived to this date. They appear to have been an intelligent species too, so perhaps they may have changed their eating and dwelling styles accordingly.

The techniques of surviving might have embedded in their brains or they may have been passed on to the younger generation through teaching.

Neanderthals also coexisted with humans, and there can be a possibility also that they evolved themselves to reside in harsh Himalayan conditions, but as their physical stature was similar or smaller than average recent humans, and with no fur, it's very less likely that they are the abominable snowman.

Another candidate can be *Homo floresiensis* which existed about 50,000 years ago, coexisting with humans, but as they were very much shorter then us and were around three feet tall with no fur, so it's also highly unlikely that they could be

 **Gigantopithecus blacki**
 Homo sapiens

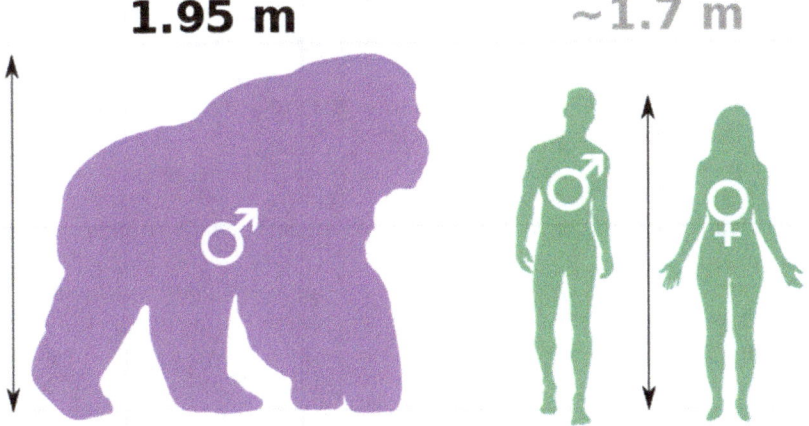

1.95 m   ~**1.7 m**

the abominable snowman. But still, there could be a very slight chance that due to some kind of extreme natural selection or genetic mutation they have grown fur all over the body or may have increased intelligence and by that intellect they may have started making fur blankets out of the fur of bears or other animals and hence every hair sample found from them may have came from those fur blankets so DNA analysis showed it as bear's or other animal's fur or hair.

Anything could be possible.

But the most believable theory is that *Gigantopithecus blacki* has evolved itself to cope with harsh climate of Himalayas.

Animals living in cold climates have thicker coats of fur. Some animals grow fur of different colours to better blend with the surroundings; an example being the white tailed deer, it grows greyish fur coat to blend in well with the colours of tree trunks and snow. Some accumulate fat and body mass to get better insulation and cope with cold winter. They also need to have a low surface area to volume ratio to prevent undue heat loss. Some animals start huddling together in groups to retain warmth. Some go into hibernation to reduce food consumption and burning of body mass.

As we know, *Gigantopithecus* were already big with huge amount of body mass with thick fur so they were already equipped with helpful features to survive in cold and harsh environment and as a slightly intelligent creatures it perhaps stayed in a group and hence retaining the heat among themselves.

These features shows they were already a good survivor so it's not much difficult for them to evolve themselves into a better survivor in this era.

They may have started hibernation or have evolved a process to suppress their metabolic needs so that they can save their energy and to carry out both as they have to restrict their movement and perhaps that's why it's hard to track or see them outside.

There is another feature which may have helped them to survive and is perhaps still helping them to survive. This feature has been stated in all the mythology, lore and tales associated with yeti and hence perhaps making a feeble link between *Gigantopithecus blacki* and the yeti. And it is the intelligence.

We all know that most intelligent creatures on the planet are apes so it's possible that yeti is an ape and that the physical resemblance of yeti is similar to an ape. So it could be a evolved *Gigantopithecus*. *Gigantopithecus blacki* was a creature with brain capacity of approximately 300-500cc . I have hypothetically deduced it by comparing it with it's previous and recent distant relatives who had similar brain capacities. The more cubic centimetres and therefore more volume may indicate slightly more intelligence.

Australopithecus afarensis, a very distant relative of *Gigantopithecus* had cranial capacity (brain volume) of 450 CC and *Homo naledi,* an another distant relative of *Gigantopithecus* and modern humans had a cranial capacity of 465-610cc. We, the humans, have a cranial capacity of 1400-1500cc. Cranial capacity of 300-500cc is

There is also a theory that they may also posses feelings like friendship, gratitude, judgement etc because it has been found that many apes and even *bears* have the ability to judge and have feelings like friendship and gratitude.

The brain may have evolved to be bigger, more complex and better, with the passage of time due to the facing of environmental challenges. Brain evolution is an ultimate expression of neuroplasticity which is reorganisation of neural networks and regeneration of new neurons.

Our brain size, for example, increased rapidly (approximately threefold) from 800,000 - 200,000 years ago due to changes in environment which forced our brains to get stimulated. This meant that the number of neurons increased (by neuroplasticity) and also majorly due to natural selection in which those with less brain power got eliminated and more intelligent ones survived producing offspring. Every part of our brain's gyri and sulci (which are the grooves and folds of our brain containing neurons) got bigger including the frontal lobe which is associated with executive functions like self-control, planning, making strategies, reasoning, abstract thoughts etc. Temporal lobe which helps in interpreting and processing auditory and visual stimulation and helps in language recognition and also in forming new memories. Parietal lobe which helps in integration and processing of sensory signals like tactile signals. It also helps in processing spatial senses and navigation etc. Same could have happened with *Gigantopithecus*; the change in environment having affected their brains or naturally selected those individuals with more intelligence and gradually evolving

also not too bad, as with this intellect they might have identified different things easily, could have done deduction of some basic occurrences and could have made some basic tools. Their descendants, the orangutans, have - after all - been observed making tools. I inferred that they were already slightly more intelligent beings so they may have evolved to become even more intelligent right now/ If they exist, and have understood humans and their surroundings better, they would very likely be able to stay away from human eyes.

them, transforming them into more intelligent creatures and maybe the increase in efficiency in planning, reasoning, strategy and decision making, processing senses and navigation is the key which might have helped them to survive and stay away from human's reach. It is certainly a possibility.

It is also possible that the abominable snowman is a totally different creature then the one we have expected. This could be possible because due to evolution the recent creature (yeti) will have many, totally different features, looks and habits then its ancestral relatives.

There are many ways an organism can evolve and change; gene flow, genetic drift, natural selection, mutations etc. Natural selection is present whenever there is heredity, variability and struggle for resources, despite infinite potential to grow. If the individual is not well enough adapted and lacks certain features it will be extirpated by nature and other organisms or individuals with those features will survive, thrive and prevail. And with new generations, those features will be amplified, hence creating totally different and a creature with a completely new appearance than before. Genetic mutations are random changes in DNA sequences which create different proteins which becomes finally responsible for different looks, habits and features and hence responsible for creating a totally different organism then before. So any creature

might have gone through these and may have changed itself into a totally different looking creature which we call "The Abominable snowman", "Bigfoot" or "yeti" .

Although we all accept that it would be very difficult to find a yeti and some of my above points suggests this. But I think there are still many opportunities and procedures to investigate and find the yeti so we shouldn't give up hope.

Some of these procedures could involve searching for places with ample food resources and many hiding places and if there are sightings and footprints in those areas there is very much chances that yetis are dwelling there.

Then fragrant food can be placed as bait. Music or other sounds might also help in attracting them. Or, conversely, it may be that the yeti will try to flee away from that place as a result, but in both circumstances their movement can be tracked.

As an intelligent creature, they can stay still also in their respective habitats, without making any movement to avoid getting caught. So, perhaps, if we can convincingly pretend that we had left the area by restricting our movements, leaving them to think we are gone, they will surely come out to check but most probably they will be attracted by the bait.

There is also the probability that yeti are residing in places where food and hiding shelters are not enough and climate is harsh. This would be because they are already evolved to reside in such places and hence can somehow live there. This would mean that the yeti can only be investigated if we keep tracks of the footprints, sightings and unknown kind of faeces.

Increasing the usage of equipment like sound navigation and ranging, thermal sensing devices, infrared sensors can help a bit but it will still be hard even with those tools, if they hiding and residing in caves or behind trees, but equipment like this could well increase the probability of discovering the yeti.

But finally, I just want to say that it will be very exciting, exhilarating and thrilling to meet a real yeti and find more about it.

It will prove to be a new revolution in science and will open new windows of research related to biological diversity.

DISCUSSION DOCUMENT: DUTIES FOR REGIONAL REPRESENTATIVES

Ronan Coghlan

The CFZ had had regional representatives for over twenty years now. Some of them have done remarkable things, some nothing at all, and some something in between. I originally intended my first wife to manage the list of regional reps, but as history shows, that never happened.

Ever since Alison and I split up I have been intending to ask someone else to take over the job, and finally a few months ago I got around to it. Ronan Coghlan has agreed to take over the onerous task, and

has come up with a list of suggested roles for regional representatives, which I post here for public discussion.

[1] In the event of a reported sighting of a mystery animal in the representative's area, all possible data should be gathered and forwarded to CFZ. Likewise, news of further developments should be sent on as they occur.

[2] Representatives should try to discover if there were any sightings or other anomalous events in their areas in the past, but should only send on stories of UFOs or ghosts if they consider them important, as otherwise their is the danger of CFZ being swamped.

[3] Representatives should, if possible, look into local folklore to discover if stories of anomalous events in the area occur. Liaisons should be initiated with the Bird, Butterfly and Conservation Officer in their areas where possible. They should, in addition, try to gather an archive of Fortean zoological material from their local studies libraries.

[4] Representatives should initiate liaisons with groups dealing with anomalies and nature in the area, provided they consider them and their personnel suitable.

[5] Representatives should have the option of offering sales of books to local bookshops. However, some might find this distasteful and so this should not be regarded as an actual representatives' duty.

ANIMALS IN THE BLOOD
The Ken Smith Story

A Biography of Gerald Durrell's Right-Hand Man

RUSSELL TOFTS

- **Publisher** : Bartlett Society (29 April 2012)
- **Language**: : English
- **Hardcover** : 284 pages
- **ISBN-10** : 0953158845
- **ISBN-13** : 978-0953158843

My family and I went on holiday to the Channel Islands in 1966, and the thing that I was most excited about was visiting Jersey zoo, because I was a Gerald Durrell fanboy par excellence.

And for months I had been telling anyone who would listen that visiting Jersey Zoo would be a unique experience, because Gerald Durrell had invented a new kind of zoo and that it was going to be something completely different to anything that we had experienced before.

Well it wasn't. It was crap.

And for many years it remained one of the biggest disappointments in my young life. My first impressions of it were that it looked home-made, and that some of the cages looked less impressive than the big hen coops where Gran and Grandad kept their chickens. I had been particularly looking forward to seeing the tuataras, but all there was to see was a ramshackle arrangement that looked like a badly made lean-to garage, and the pair of unique reptiles, which - whilst they looked a little like lizards were nothing of the sort - were nowhere to be seen.

I am sure that there must have been something worthwhile there, but to a downtrodden six-year-old for whom Gerald Durrell had assumed an almost Godlike status, it was a horrid shock. Six years later I read Durrell's book Catch me a Colobus which explained a little about what had happened. Durrell and his then wife Jacquie had been on a series of extended trips, either making films or catching animals, but were shocked when they came back from the extended visit to Malaya and the Antipodes which is described in Two in the Bush to find that - in Durrell's words – "my precious zoo was looking shabby and unkempt and that it was almost bankrupt". Durrell took over direct day to day management of the zoo, and the project was eventually saved.

I found out more when I read Douglas Botting's biography of Gerald Durrell many years afterwards. Apparently he had started the zoo with a bloke called Ken Smith with whom he had worked at Whipsnade, and been on several of his more high profile expeditions. It appears that Smith was Durrell's business partner, and had been manager of the zoo for its first four years of existence.

Apparently there was a massive falling out between the two men, which resulted in Smith leaving the island and moving to Devon where he started a number of zoos, most notably Exmouth, which I remember with no great fondness and which was (according to Getty Images) closed on humanitarian grounds in 1980.

On first reading Botting's book over the Christmas of 2002, I had a quick shufti online to see if I could find any corroborating evidence, and found plenty. I wrote Smith off as a bad, or at least a misguided, egg and thought

very little about him for the next fifteen years. However, about an hour before writing these words I went back online and found that in the intervening decade and a half Smith's reputation has been through some sort of sea change, mainly due to a book by Russell Tofts, which I decided that I really should get round to reading.

And boy was it an eye opener.

I already knew from Douglas Botting's biography of Durrell that he was a prodigious drinker, and an out-of-control alcoholic for much of his life, and that he could also be a bit of a dick. It appears from what Russell writes that there were faults on both sides. Gerald Durrell was indeed a bit of a dick to Ken, but – then again – they were totally different people with different mindsets and different views of what a zoo should be. To Gerry, Ken was dull and money oriented. To Ken, Gerry was an annoying hedonist with very little grasp of the realities of day-to-day existence.

But it also looks as if Ken left voluntarily rather than being pushed. And then that – for the rest of his life – Gerry tried to expunge him from his personal history.

But that is not the weirdest thing. It turns out that in the years leading up to the Second World War, Ken Smith worked as an insurance agent for the very same company that both my grandfathers worked for, and in the same area of the Cotswolds. They must've known each other. And then, even more peculiarly, I found myself quoted in the pages of the book, in the very same passage that I had been hoping to expand. The universe truly is a very strange place.

The book also tells us quite a lot about the way that reputations are made and lost both before, and in these days when the information superhighway has blazed its inimitable way through all of our lives, and also underlines the fact that who are the good guys and who are the bad guys is very much a matter of historical perspective.

The sort of zoos that Ken Smith ran are very much a thing of the past, as our society moves more and more towards the idea that it is unethical to think of animals as merely items for human entertainment. But we had not yet got to the ideal that I espouse shortly before an irreparable falling out with the proprietors of just one such tourist orientated zoo. I told them that, to my mind, a zoo should be a temple at which one worships at the altar of Mother Nature. In his masterpiece, 'Kim', Rudyard Kipling has the title character describe the museum at Lucknow as "The Wonder House", and that is how I believe museums, zoos and public aquaria should be, and so often are not.

Despite the fact that Durrell himself had, if not libelled then come very close to libelling, Ken Smith, he was very much a man of his time, and should not be judged by 21st century standards. There is too much of that going on these days; something with which I vehemently disagree. Ken Smith played a very important part in the history of British zoos, and has sadly been maligned since his death. This book does much to redress the balance, and I am very grateful to Russell Tofts for having written it.

- **Publisher**   :Aleph, India (11 April 2013)
- **Language** : English
- **Hardcover** : 304 pages
- **ISBN-10** : 9382277552
- **ISBN-13** : 978-9382277552

In the\ sixteenth century, Dutch traveller Jan Linschoten noted the absence of lions throughout the Indian subcontinent. Two hundred years later, echoing similar comments made by various hunters and observers of Indian wildlife, the British shikari and writer, Captain Thomas Williamson, emphatically declared: There are no lions in Hindustan. Much the same was said about the cheetah in the region.

These observations piqued the interest of well-known naturalist Valmik Thapar. After an enormous amount of research and study he now believes that, contrary to existing scientific theory, neither of these animals were indigenous to the Indian subcontinent. Remarking on the lack of accounts of encounters with these animals as opposed to the tiger and the leopard which are extensively documented as well as inconclusive genetic studies, he argues that, over the centuries, the lion and cheetah were brought into the country from Persia and Africa by royalty, either as tributes or to populate their hunting parks and menageries.

Enlisting the help of renowned historian, Romila Thapar who analyses historical accounts and representations of the lion in early India and scholar, Yusuf Ansari who looks back at the lives of the Mughals and their famed hunts to further validate his theory, Valmik Thapar concludes at the end of this thought-provoking book that the Indian lion and the Indian cheetah were, in fact, exotic imports, and not indigenous subspecies. Tracing the history of the lion and the cheetah for over 5,000 years, and substantiated with pictorial evidence, Exotic Aliens is a pioneering work that could turn field biology on its head.

As regular viewers of my weekly webTV show, On The Track, will know, I have recently fallen down a rabbit hole whereby various people have been claiming that the Asian cheetah (Acinonyx jubatus venaticus) is still a resident of India and Bangladesh. This is pretty much nonsense, because accepted wisdom has it that the Asian subspecies of the cheetah was extirpated from the Indian subcontinent during the first half of the 20th century, with the final three specimens being shot by a trigger happy Maharajah in 1949.

However, a string of recent news items from Indian newspapers that have been picked up by Google News Alerts have claimed that cheetahs are alive and well and living in Bangladesh. It intrigued me to find that each of these news items was accompanied by photographs of jaguars - which are, of course, a species that has never been found outside the New World - and I hypothesise that for some reason the term 'cheetah' seemed to be usable by the journalist responsible for these stories (I have no idea whether they are native Urdu, Hindi, or speakers of some other language) for any one of the spotted big cats.

One of the regular viewers of On The Track suggested that I read this book, and as is so often the case - when studying anything to do with mystery animals - it gave us more questions than answers. First of all, however, it is interesting to know that on page 575, the authors note:

"References in Sanskrit texts are also not very clear and terms such as dvipi and chitraka (spotted), can refer to leopards and panthers as well. The first is used in earlier texts and the second in later works."

The authors also note that, in the Indian language Urdu, the word 'cheeta' is used for both the leopard and the cheetah, which does satisfactorily explain why there seems to be so much confusion between cheetahs and leopards in modern Indian journalism. It doesn't, however, explain why in both of the recent occurrences that I have come across, the image used to illustrate the offending article is of a South American creature which has never lived in India.

But none of this is really what this book is about, and I found myself – whilst searching for the truth at the bottom of my own rabbit hole – totally engrossed in the intellectual argument that these authors present.

A week ago, if you had asked me, I would have said that the Asian cheetah was once upon a time a resident of large swathes of the Indian subcontinent, and indeed had a range which stretched from Hindustan all the way to Western Asia, and possibly even as far as the Mediterranean. And I would furthermore have said that the Asian lion (Panthera leo persica) had in historic times been found everywhere from Greece to India, having been extirpated from Palestine during or soon after the First World War and Arabia, Iran and Iraq in the 1940s or '50s, and that this beleaguered subspecies was now only to be found in the Gir Forest in Western India. Well, as is so often the case, it appears that everything I thought I knew about the subject is quite possibly wrong.

I was discussing this book with Carl Marshall the other evening, and said how interesting it was to see somebody utilising cryptozoological methodology, for once not to prove the existence of a disputed creature but to disprove it. Carl suggested that this was an example of anti-cryptozoology, but I disagree and told him so.

The whole essence of cryptozoology, and by extrapolation, Fortean zoology, is not merely to find new species of hitherto only ethnoknown animal, it is to find the truth behind the world's great animal mysteries. And, I had not realised until a few days ago quite how great this particular animal mystery actually is. Indeed, I didn't even realise that it was a mystery.

To deal with cheetahs, first of all, whilst quite sensibly not taking a pot shot at the veracity of the Asian subspecies of the cheetah itself. As far as we are aware, these massively rare animals only exist in tiny numbers in a couple of populations in Central Iran and possibly a population in Southern Turkmenistan.

However, the authors claim that there is no cultural residence for the cheetah in the Indian subcontinent before the Middle Ages when coursing with cheetahs as a beast of the foot became a popular pastime amongst the Indian ruling classes, and huge numbers of cheetahs were imported from Africa to live in the menageries of the kings, princes, and princelings that were found all across the Indian subcontinent. I hadn't realised until now how every ruler worth his salt had a private menagerie, and quite often the menagerie contained lions or cheetahs. Indeed, there was a massive trade in wild animals from Africa to India and even further east.

And the authors posit that "wild" cheetahs in India were not the eastern most members of an Asian subspecies but were actually menagerie animals that had escaped, and presumably bred.

The history of the Indian lion is even more complicated. I remember reading in an online paper of the subject that genetic tests on various zoo specimens of the Asian lion had – within the past twenty years or so – proven that the Asian lion bloodline, which is, as far as I'm aware, entirely made up of animals from the Gir Forest population, had disturbing amounts of African lion DNA within it.

The inference was that the breeding programmes in zoos had not been as continent as they should have been, but this book presents a fascinating, and really rather disturbing, alternative explanation.

Again, the authors are not disputing that the Asian lion was once found in the Middle East, but the wider claims of the book are that, as the Muslim sphere of influence expanded quite rapidly eastwards, the iconography of the lion as a royal beast, or as the king of beasts, travelled with it.

And so, over a thousand years ago, every Indian ruler worth his salt wanted lions in his menagerie, both to exhibit and to slaughter as part of a coming-of-age ceremony. I hadn't realised that the ritual slaughter in arenas amongst various Indian civilisations was as advanced and socially complex as the better-known slaughter of animals and people by the Romans.

And, I had not realised that such a complex system of commerce had grown up in order to facilitate this.

Again, using the cryptozoologically accredited methodology of checking the art and culture of an area as an invaluable aid to discovering the existence or otherwise of animals, it appears that when you go immediately to the west of India, there is no evidence that there ever had been populations of Asian lions in what is present day Pakistan, and indeed the nearest truly attestable populations were in Afghanistan and Iraq.

And it appears that there is a very good argument to be made, suggesting that the lions which were "wild" in India in historic times and are still to be found in the forest of Gir in the present day, are a bloody awful mishmash of Asian and African bloodlines, and therefore the amount of effort that has been put into conserving the Gir Forest population is arguably completely pointless.

It is generally accepted that there was a huge genetic bottleneck in the cheetah bloodline approximately 100,000 years ago and another one approximately 12,000 years ago. Which means that the resulting level of genetic variation is considerably smaller than in most living species. Even so, it is still claimed that the Asiatic cheetah diverged from the cheetah population in Africa between 32,000-67,000 years ago, and that there is enough evidence to suggest that geographic isolation has cause subspeciation.

However, having read this book I, for one, am perfectly prepared to accept that Asian subspecies never made it as far as the Indian subcontinent, unless it was with humanity as a vector.

The current status of the Asian lion is even more ironic. After finishing this book I did a bit of digging and I found out that, a few years ago, P. l. persica was determined to be a junior synonym of *P. l. leo* and so, if this is indeed the case, despite the Indian government - and before then, the British colonial authorities - doing their utmost best to preserve the remaining Indian lions in the Gir Forest, it was not only a waste of time because the population is so bastardised, but that even the bona fide Asian lions (or their genetic remnants) are actually exactly the same subspecies as those found in the northern half of Africa.

Funny old world, innit?

- **ASIN** : B0BGQ5L15L
- **Publisher** : Independently published (28 Sept. 2022)
- **Language** : English
- **Paperback** : 110 pages
  ISBN-13 : 979-8355119430

And now time for a confession: if it hadn't been for this magazine, edited by our cover artist this issue, we would never have resumed doing hard copy editions of this one, and I feel churlish that we haven't got more room for it.

However, squeezing everything into a smaller space as possible, this is a damn good magazine, covering Bigfoot research in all its confusing glory and, unusually, presenting both sides of every story. Steve deserves to be seriously congratulated as well as provoking me to get my finger out and sit back in the editorial chair of this magazine.

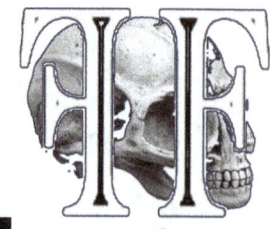

CFZ PRESS

CFZ CLASSICS

MYSTERY ANIMALS OF THE BRITISH ISLES

THE WORLD'S WEIRDEST PUBLISHING GROUP

We publish a lot of books. Indeed, I think that we could quite easily claim to be the world's foremost publishers of books about Fortean Zoology and allied disciplines, and our Fortean Words imprint is doing a great job in producing books on other non-zoological esoterica. However, I feel that it would be unethical to review our own titles. So here, to end this edition of *Animals & Men*, is a brief look at the books we have put out so far this year.

- **Publisher :** cfz (1 Aug. 2021)
- **Language :** English
- **Paperback :** 306 pages
- **ISBN-10 :** 190948864X
- **ISBN-13 :** 978-1909488649
- **Dimensions :** 15.6 x 2.11 x 23.39 cm

In 1957 Soviet historian and social scientist Boris Porshnev, inspired by the reports of the Yeti in the Himalayas, became interested in the possibility of similar creatures in the area of Europe and Asia then controlled by the Soviet Union. He was given permission by the Soviet Academy of Sciences to establish a Commission to examine the whole question of the 'Snowman'. After he wrote an article in *Pravda* he received over a thousand reports from all over the Soviet Union, giving a consistent picture of a wild creature, more closely related to human beings than any known species, surviving in mountainous areas all over Asia. An expedition to the Pamirs of Tajikistan was organised in 1958 to follow up the most promising reports. Unfortunately, more powerful figures in the scientific establishment subverted the original purpose of the expedition and it produced little result. From then on Porshnev's position declined. His theory that Asian wildman reports could be explained by surviving Neanderthals was attacked, in one case in terms that doubted his sanity. The defence he wrote could not be published in Moscow, and had to appear in a Kazakhstan literary magazine.

In 1963 he produced a book summarising the evidence the Commission had received, studies from other parts of the world, and further evidence from history. He built up on this basis a consistent picture of the creature and discussed its possible relation to Neanderthal man. The book was never actually completed, but 180 copies of a preliminary version were circulated to colleagues in Moscow. The book then disappeared for well over half a century. With the assistance of Porshnev's family the manuscript has now reached the West and is published here in an English translation with the addition of notes, maps, illustrations and an index. This book casts a wholly new light on the Yeti, Bigfoot and the possible survival of human ancestors into the present day.

Buy the book at a special low price today.

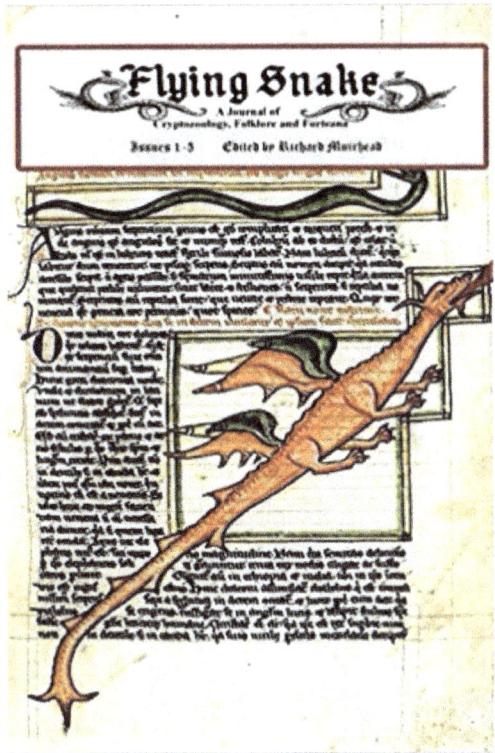

- **Paperback**: 392 pages
- **Publisher**: cfz (22 Mar. 2019)
- **Language**: English
- **ISBN-10**: 1909488585
- **ISBN-13**: 978-1909488588
- **Product Dimensions**: 14.8 x 2.1 x 21 cm

Richard Muirhead is probably the best researcher that I know.

He has, what I am sure Charlie Fort would have called, a 'Wild Talent' in that you can let him loose in any library or archive, and with the innate powers of a truffle-hound he will extract all sorts of pieces of arcane knowledge that one would nev-

cane knowledge that one would never have suspected to be there.

For example, about a decade ago, he phoned me telling me he was going to look in the archive of the Hong Kong Yacht Club.

"Whatever for?" I asked and received a completely non-committal reply.

That evening, he telephoned me to tell me that he had found a hitherto unsuspected sea-serpent report.

That's just the kind of guy he is.

Some years ago, he started his own magazine, *Flying Snake*, which has proven to be a wildly eclectic and thematically diverse publication that, every few months, continues to delight and amuse.

Here, in book form, is the first volume of the collected editions.

Enjoy,

- **Paperback**: 262 pages
- **Publisher**: Fortean Fiction
- **ISBN-10**: 1909488569
- **ISBN-13**: 978-1909488564

Some years ago, I wrote a book called The Song of Panne, which told the story of how my dear, late wife Corinna and I ended up having a hairy humanoid forest Godling (to steal Kipling's nomenclature) living in the airing cupboard in what used to be my father's dressing room. You can take it as fiction if you like, or you can believe every word I say.

Believe it or not, it truly doesn't matter to me one way or another.

This book continues the story roughly from where it left off, although the vast majority of it tells the story of a few days in the Autumn of 2015, and explains some of the things left hanging in the first book. I would like to think that you can read this book as a standalone volume, although - as an impecunious author with family, animals and forest Godling to support - I could always do with the cash if you would like to buy the previous book.

Although some of the characters in the book are real (whatever that means), in order to protect my kneecaps, it is probably better if you assume that everybody here except for me and Corinna is completely made up.

I would warn you that if you are of a nervous disposition, or easily offended, you will find parts of this book both offensive and upsetting.

There is sex, violence, drug abuse, occultism, pornography, firearms, politics, religion, and not a little sociology. But there is also love, kindness, faith, and redemption.

- **Page Count**:256
- **ISBN/SKU**:9781909488571
- **ISBN** Complete:9781909488571
- **Publication Date**: 10/22/2018

Have you ever wondered what lurks out there in the deep, dark woods of the North?

This book presents a choice selection of monstrous beings and fabulous creatures from Greenland, across the North Atlantic to Scandinavia, and the Baltic States. Meet the Giant Gull of Greenland, the terrifying Skrimsl of the Icelandic wilderness, the trolls that stomp through the impenetrable Nordic woods, and the puk-dragons guarding the houses on the Baltic Coast.

This is the book where zoologist Lars Thomas tries to make sense of all the very strange beings that live in these Curious Countries.

- **Language** : English
- **Paperback** : 348 pages
- **ISBN-10** : 1909488631
- **ISBN-13** : 978-1909488632
- **Dimensions** : 15.6 x 1.83 x 23.39 cm

Jonathan Downes is the director of the Centre for Fortean Zoology, and he is undeniably one of the best known cryptozoologists in the English-speaking world. Jon has dedicated his life to searching the world for mystery animals and trying to make sense of the mysteries of Mother Nature. These preoccupations started over half a century ago, in a place which - for all intents and purposes - no longer exists and which contemporary readers may struggle to comprehend.

Between 1961 and 1971 the Downes family lived in Hong Kong, which is where, surrounded by exotic animals and oriental folklore, he first fell in love with the natural world and the mysteries therein.

This is not only the story of his early life, but also the story of one of the last generations of children brought up in the aegis of the British Empire, when it was still a global entity upon which the sun had not set. The author takes an unprejudiced look at the last decades of British rule in the Orient, through the eyes of the child of a senior member of HM Overseas Civil Service.

This book has taken a lifetime to write and it examines his Colonial world at face value, neither exaggerating or shying away from the truth. It is a book about love, hate, mental illness, prejudice, duty and compassion. As such it can be frightening, touching and confusing. It will make you angry, happy and sad, but above all it is about a whole pantheon of wonderful animals. Six decades later the author has not lost his childlike wonder at these creatures and at the magnificence of the world around him.

2022/2023 YEARBOOK

Edited and Compiled by
Jonathan Downes
Richard Muirhead

- **Publisher** : cfz (1 April 2022)

- **Language** : English

- **Paperback** : 260 pages

- **ISBN-10** : 1909488658

- **ISBN-13** : 978-1909488656

- **Dimensions** : 15.6 x 1.8 x 23.39 cm

The Centre for Fortean Zoology (CFZ) is a professional and scientific organisation dedicated to cryptozoology: The study of unknown animals and allied disciplines. Since 1992, we have carried out extensive research into mystery animals and animal mysteries around the globe.

We produce a weekly WebTV show called On The Track (OTT), which covers Cryptozoology, Natural History and Green Issues, all mixed with a little light (and often peculiar) comedy. We also operate our own publishing house, producing both magazines and books on subjects that would otherwise not see the light of day

The Centre For Fortean Zoology Yearbook is a collection of papers and essays too long and detailed for publication in the CFZ Journal Animals & Men. With contributions from both well-known researchers, and relative newcomers to the field, the Yearbook provides a forum where new theories can be expounded, and work on little-known cryptids discussed.

CONTENTS

- **Publisher** : cfz (5 May 2022)
- **Language** : English
- **Paperback** : 130 pages
- **ISBN-10** : 1909488666
- **ISBN-13** : 978-1909488663
- **Dimensions** : 15.6 x 0.89 x 23.39 cm

This is a remarkable book told from a unique perspective. Damon Corrie is a hereditary chief of the Eagle Clan, of the Arawak Tribe based mostly in Guyana. He has made a lifelong study of the mystery animals and animal folklore of his people, and we believe that this is the first time that these remarkable accounts have been collected together in a single volume.

On top of that, the CFZ's very own Richard Freeman has added a number of appendices describing the expedition that he and Damon went on back in 2007.

We heartily recommend this new volume to anybody interested in the mysteries of South America, and the mythology which shapes the people who still live there.

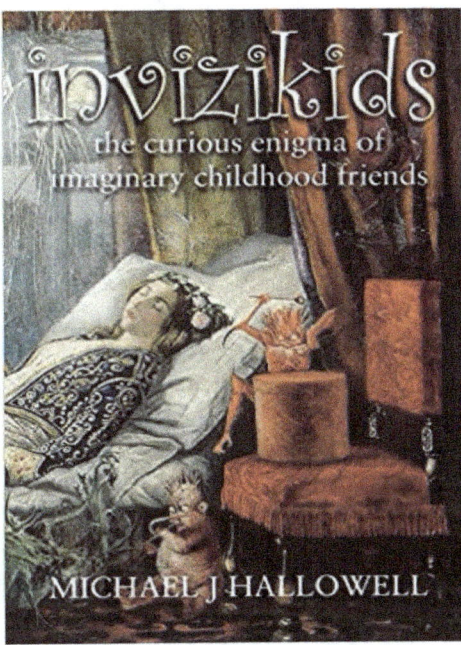

- **Publisher** : Fortean Words (11 Nov. 2022)
- **Language** : English
- **Paperback** : 166 pages
- **ISBN-10** : 1909488674
- **ISBN-13** : 978-1909488670
- **Dimensions** : 15.6 x 0.89 x 23.39 cm

This is a fantastic book. It is about the imaginary friends which so many children have. I had one when I was a child, and probably so did you. Writing in 2015, eight years after this book was first published Lawrence Kutner, Ph.D. says:

"Imaginary friends are an integral part of many children's lives. They provide comfort in times of stress, companionship when they're lonely, someone to boss around when they feel powerless, and someone to blame for the broken lamp in the living room. Most important, an imaginary companion is a tool young children use to help them make sense of the adult world."

These friendships take place in a weird sort of ur-space that is neither pure imagination or actual reality (whatever that is) and are important not just because they are practically universal, but because of their implications for those of us who study the noosphere.

I have known Mike Hallowell for well over 20 years, and have always been impressed by his books and articles. He has the sort of enquiring polymathematical mind which I admire, as he sets his talents into investigating a wide range of different arcane subjects. But this enquiry into the true nature of childhood imaginary friends may well prove to be the most important thing he has ever written.

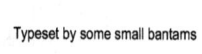
Typeset by some small bantams